Rivers

CURRICULUM GUIDE

BIOLOGY

One of a series of six rivers-based units written by teachers participating in the Rivers Curriculum Project funded by the National Science Foundation

Dr. Robert Williams, Project Director
Southern Illinois University at Edwardsville (SIUE)
Cynthia Bidlack, Project Coordinator, SIUE

Lead Author:
Dr. Robert Williams, SIUE

Contributing Authors:
Dr. Dan Bean, St. Michael's College
Bill Beckman, East Peoria High School
Dr. Taylor Delaney, Principia College
Bill Donato, Woodstock High School
Don Dickey, Civic Memorial, Bethalto High School
Janet Franklin, Meredosia-Chambersburg High School
Jim Gager, Moline High School
Bill Henske, Southern Illinois University at Edwardsville
Dr. Art Hessler, St. Michael's College
Marvin Mondy, Lewis and Clark Community College
Kathy Newsom, Sherrard High School
Al Schuitema, Fulton High School

Dale Seymour Publications®
Parsippany, New Jersey

Managing Editor: Catherine Anderson

Project Editor: Christine Freeman

Production Coordinator: Joe Conte

Design Manager: Jeff Kelly

Text and Cover Design: Christy Butterfield

Cover Photograph: Nicholas Pavloff

Illustrations: Joe Conte

Dale Seymour Publications®
An imprint of Pearson Learning
299 Jefferson Road, P.O. Box 480
Parsippany, New Jersey 07054-0480
www.pearsonlearning.com
1-800-321-3106

ISBN 0-201-49369-1
2 3 4 5 6 7 8 9 10-ML-01 00

Printed on acid-free, 85% recycled paper (15% post-consumer), using soy-based ink.

This book is printed on recycled paper.

Acknowledgments

Rivers Project Curriculum

We cannot list all the teachers, scholars, and friends to whom we are indebted for the development of the Rivers Project curriculum units, but we want to mention some of these very important people.

To start, without the help and guidance of Mark Mitchell and Bill Stapp, GREEN Project at the University of Michigan, Ann Arbor, in the very early stages of the Rivers Project, we might not be where we are today. The same thank-you goes to Tanner Girard, Illinois Pollution Control Board Judge and former professor at Principia College in Elsah, Illinois, who has been helpful in so many ways. We owe so much to Don Humphreys and Ivo Lindauer, our National Science Foundation Program directors, who have believed in and supported the curriculum development since its inception and throughout the long and tedious job of writing, piloting, rewriting, and field testing. We offer special thanks to the Illinois State Board of Education and the Illinois Higher Board of Education for beginning the project and for offering continuing support for Dwight D. Eisenhower teacher training.

No way exists for us to thank sufficiently each of the curriculum unit writers for the hours and hours of personal time devoted to their specific unit. Not only have they been writers but also trainers for new teachers entering the program, during both the school year and the summer training sessions.

We want to thank all the curriculum unit users for revision suggestions given to the writers. We are especially grateful to them for taking time away from their traditional classroom setting to give time to their students out on the river for hands-on activities.

To the university professors from across the country who read the units in their area and provided professional opinions and suggestions for improvement, we are sincerely thankful.

Finally, to Pat, Jack, Bill, Michele, and all the many university student workers and graduate assistants who have come and gone over the course of the last four years, what would we have done without you?

Dr. Robert Williams, Project Director
Cindy Bidlack, Project Coordinator

Cynthia Lee (Cindy) Bidlack died of cancer shortly after her 46th birthday. She gave so much to this project. For all her efforts, we can only say thanks. We remember her often and miss her much.

Rivers Biology

The writing of *Rivers Biology* was a cooperative effort. Some joined the production for a little while, adding ideas, reviewing, or teaching the materials as they helped to form the unit. The effort was complicated, but the ownership is spread across many miles and is embedded in the minds and hearts of countless students who served as practitioners. Christine, you have prodded us all, but we are proud of our efforts. To all of those who study rivers, our hope is that these lessons will provide a basis for you and your students to learn and explore life at the river. The rivers of the world need the combined efforts of us all; and we need those rivers to enjoy and to use.

CONTENTS

The Rivers Project Curriculum

When many of us think of rivers, we picture the fun times we have around them—boating, swimming, frolicking, watching wildlife, just enjoying being where land meets water. We may also appreciate the benefits that rivers give to us, such as drinking water, plentiful food sources, recreation, electric power, and an efficient means of transportation. While tallying up all these good things about rivers, however, we may grow concerned about threats to the health of our rivers.

Rivers and streams confront many forms of pollution—industrial waste, acid rain, sewage spills, and thermal pollution. Fortunately, scientists, environmentalists, and the public have instigated regulatory and technical changes that seem to be reducing some of these risks. What kinds of river pollution, and effects of pollution, need further attention? What other threats do rivers and streams face?

Scientists increasingly tell us that the main threat to America's rivers today comes not from pollution, but the physical and biological transformation of rivers and their watersheds. As rivers are altered to provide water transportation, generate power, reduce flood hazards, and provide water for farms, cities, and industries, their physical, chemical, and biological processes are damaged or destroyed. The loss of riverside and aquatic habitat has led to the decline or extinction of more than one-third of North America's fish species and an even higher proportion of its native mussel species.

Healthy river systems are incredibly dynamic. As nutrients, sediments, and organisms are transported downstream, water and organic materials are constantly added to the mix. Most of these materials come from the surrounding terrestrial system, with the land-water boundary, known as the "riparian zone," acting as a critical valve or filter that regulates the exchange. Riparian zones and their associated wetlands also act as natural sponges, absorbing and filtering polluted floodwaters over time. Where the banks of streams are cleared, straightened, and replaced with rocks or concrete to reduce flooding, the ability of associated wetlands and floodplains to control and filter runoff, provide habitat, and add nutrients is lost.

When rivers flood, they alter the shape of the stream, scouring new channels, inundating riverside land, depositing sediments, and building new banks and beaches. These functions are as important to healthy river ecosystems as natural fires are to healthy prairies and forests. For many fish species, this flood "pulse," called the "natural hydrograph" by scientists, not only triggers spawning and migration but also allows fish to reach seasonally inundated floodplain nurseries and spawning habitats.

Scientists have made significant progress in understanding how changing natural hydrologic cycles has contributed to the destruction of aquatic ecosystems. Numerous local communities have taken the lead in adopting cost-effective storm water, floodplain, and water-supply management programs that utilize natural hydrologic processes. Sometimes, however, federal and state agencies want to alter the hydrology and other physical characteristics of rivers and their watersheds. The debate on such issues ranges down many river corridors.

In order for the youth of our country to become informed participants in the political process, they must have a solid background in environmental issues. The school systems must, therefore, educate students about the nature of the environment. Through the study of rivers, not only does a concern for water become important to the students, but other issues begin to gain this same importance. And for our country, and our world, this can only be one giant step forward.

The Rivers Project curriculum is the end product of four years of environmental commitment by hundreds of high-school teachers in the United States and Canada. Because of the growing importance of environmental issues, teachers through whose towns a river or stream flows have sought to integrate water and river studies into their traditional content courses.

USING THE RIVERS CURRICULUM

When students visit a river or stream and become actively involved in observing, measuring, testing, and writing about that waterway, they quickly develop a sense of ownership toward that river or stream. They also tend to develop a broader understanding of the value of the academic discipline that has brought them to the water's edge. Toward these goals, the Rivers Project units were developed by teachers for use in science (chemistry, biology, earth science), geography, language arts, and mathematics classes. The Rivers Project curriculum prepares students to perform field investigations, with the primary laboratory being a local river or stream. Some activities are to be used in the classroom, focusing on preparation for the field experience, which may be one long trip or several shorter ones. Extensions of the curriculum units encourage teachers to make use of other kinds of field trips, to invite guest speakers to the school, and to make contact with local, state, and federal agencies for outside resources.

In a number of states, governmental agencies are using in their monitoring processes data collected by Rivers Project students. Such contact with state agencies is a vital component of the recommended scientific activities stated in the curricula for these units. As students experience hands-on learning activities that result in river data of real scientific or cultural value, affecting real-world issues they care about, they are motivated to learn even more.

The Rivers Project curriculum is not intended to be used as a textbook, but as a set of supplementary materials designed to enhance your existing program or to establish a basis for river study in your school. The materials involve students in a natural environment—the river—through a series of hands-on activities conducted through field-based study.

When using these units, let your imagination and your creativity run wild. The materials can prove a valuable addition to your traditional teaching. *Remember, what your students feel and touch as part of their river experiences will stay with them for the rest of their lives. Make it memorable!*

Rivers Project Curriculum Units

The river is the common strand weaving the units together into the interdisciplinary curricula you are about to use. This brief synopsis and the underlying connection of each curriculum unit to the other will aid in understanding how the Rivers Project units may be taught. You may use these units independently of each other, or you may combine them in an interdisciplinary approach, especially a team-teaching venture.

Rivers Chemistry defines water quality and guides students in basic data collection. Water-quality kits, which are readily available and easy to use, make conducting the tests a relatively uncomplicated task.

Rivers Biology focuses on stream-monitoring programs and the study of benthic macroinvertebrates. Living organisms in a river, stream, or lake are easily captured or documented. Their existence and numbers provide data for comparison with those of the chemical unit. Biological diversity for a water environment may change as the water quality improves or decreases.

Rivers Earth Science evaluates the physical features of a river system that provide clues to understanding the historical development within a local area. Students better understand the impact of the river drainage system on water quality when their study also factors in soil, slope, and flow. As scholars in the twenty-first century study the effects of agriculture, development, and transportation on the river and water resources, they will view the geology of an area with increasing attentiveness.

Rivers Geography enables students to develop a sense of the environmental impact of people occupying and organizing themselves along rivers. Study of the geography of the river as it relates to location, place, movement, region, and human-environment interactions along its banks gives form and reason to human migration and development. The river becomes a lab for an ever-changing society.

Rivers Language Arts focuses on the skills students will use as they investigate and write about their study of the river. Lessons include technical writing for scientific reporting, interviewing, research techniques for exploring local history, political letter writing, poetry, and other forms of creative writing. Samples of, and references to, exemplary fiction and nonfiction written about rivers are included. Teachers of every discipline can use this unit to give voices to the discoveries and ideas garnered by their students through river study.

Rivers Mathematics provides real-life application of mathematical processes and skills, using data gathered during field studies and from reference sources. Topics range from measurement and working with percentages to standard deviations. This material specifically teaches the mathematical skills required for the science units in the Rivers Project curriculum.

Scheduling and Team-Teaching Options

Each science unit, and the geography unit, can be used as a freestanding unit involving about one month of consistent focus. Alternatively, any of these units can be integrated into the regular activities of the class throughout a longer period. The language arts and mathematics units can be used in language arts or mathematics classes to support and expand river activities being done in the science classes. Science and environmental education teachers who wish to broaden student experience in other areas, or to add an interdisciplinary dimension to their curriculum, can also use these units as support material.

Because each unit can stand alone, a few units do have some duplication of topics. For example, because of its importance, analysis of fecal coliform is included in both the chemistry and the biology units. If teachers in both disciplines are using the Rivers Project curriculum, such testing may be performed in just the biology class, with the results shared with the chemistry class.

A single teacher in a school may use a Rivers Project unit, or several teachers may work as a multidisciplinary team. In many schools, what began as a single unit has grown over the course of several years into a schoolwide project as more teachers and students have become involved and the school has acquired more materials and equipment for river study.

Using Your River or Stream in This Curriculum

Teachers and students can use the Rivers Project units on any river or stream anywhere in the world. The constants are the water tests and educational studies that student perform on the river. Each river or stream presents a unique set of parameters for collecting and studying. Investigators approach bigger rivers very differently from smaller streams or shallow rivers. A cold mountain river displays vastly different flora and fauna from the slow, sluggish coastal river. Factors such as population density, natural and modified drainage patterns, and climate vary not only from river to river but from one point along a river to another. The study of each river and its corresponding watershed, therefore, has special attributes that cannot be dealt with in generic materials such as these.

You and your students will want to build a library of handbooks, field guides, and other support material appropriate to your specific area. The best sources for this information are local agencies that monitor and study the waterway. Soil and water conservation district offices, state and local conservation departments, environmental protection or water agencies, and local universities are good places to begin. Most state agencies have a number of relevant free publications. While building the library, locate a person who can advise the class in the field. Many schools have found locally based state employees who are willing to provide content support, sometimes equipment, and even to join the class on water-monitoring days. Once your class has a mentor, river study often becomes easier and more meaningful for you and your students. More than likely, the data collected by the students will also be important to the mentor.

ABOUT THE RIVERS PROJECT

The Rivers Project curriculum in these six units started with teachers who loved teaching and loved the rivers in their communities. They banded together to spread their enthusiasm, knowledge, and learning tools among students and other teachers.

History of the Rivers Project

The Rivers Project began as the Illinois Rivers Project in February 1990 as a pilot program involving eight high schools along the Mississippi and lower Illinois Rivers. With scientific literacy as the ultimate goal of the Project, students collected and analyzed water samples from test sites along both rivers. The study of the rivers was extended to include historical, social, and economic implications of the state of the rivers, thus involving students from classes across the curricular areas of science, social studies, and English.

SOILED NET, a telecommunication network linking the participating schools with each other and with Project headquarters, provided a technological framework for many of the Project's activities.

In December 1991, the Illinois Rivers Project received a grant from the National Science Foundation to develop a formal Rivers Project curriculum. The resultant units, in chemistry, biology, earth science, geography, language arts, and mathematics are applicable to any river in the world. To further the Rivers Project, Southern Illinois University at Edwardsville hosts a training session each summer. At these programs, teachers who use Rivers Project materials serve as mentors for participating teachers in classroom and field-study sessions.

Goals of the Rivers Project Curriculum

The three principle goals of the Rivers Project curriculum are:

1. to increase students' knowledge and understanding of important issues and concepts related to the river.
2. to prepare students with the necessary skills to properly investigate and report relevant information regarding the river.
3. to inspire students to take action to resolve problems that contribute to the overall deterioration of the natural beauty and functions of the river.

Project Funding

The Rivers Project began with funding from the Illinois State Board of Education through a Scientific Literacy Grant. Subsequent funding for Illinois teacher training was obtained from the Illinois Board of Higher Education, Dwight D. Eisenhower Title II funds. A grant from the U.S. Fish and Wildlife Service allowed Midwestern schools in Iowa, Minnesota, and Wisconsin to participate. The National Science Foundation, under its Materials Development Program, provided funds for the final development and testing of the Rivers Project curriculum units. Along the way, many others in the academic and business community, as well as numerous local, state, and federal agencies provided support.

TELECOMMUNICATIONS

A key component of the Rivers Project is the sharing of water-quality data and student writings among schools in the project. Though schools can communicate by mail, use of the computer networks associated with the Rivers Project provides for faster, more responsive, more directly accessible exchange. Students using these networks can not only contribute their data but also quickly access the data, writings, and queries of students at other Rivers Project schools, most notably students exploring the same river watershed.

The Rivers Project utilizes two telecommunications systems, e-mail and the World Wide Web. Either requires a computer and a modem. Because telecommunications is evolving rapidly, the telecommunications technology, data handling, and data availability for the Rivers Project will undoubtedly also change. For the most current information on how to contact the Rivers Project, telephone Southern Illinois University at Edwardsville (SIUE) at 618-692-3788 or 618-692-3065.

E-Mail

Students using a computer that has a user account can send writings to the Rivers Project via e-mail using the address "rivers@siue.edu." Other students may read these on the World Wide Web. The Rivers Project also periodically publishes collections of student writings it has received.

World Wide Web

The Rivers Project is also available via the World Wide Web. This requires web browser software and access to the Internet. The URL (Uniform Resource Locator) is "http://www.siue.edu/OSME/river." This home page contains a searchable database of all water-quality data sent in to date and also a form that can be filled out online to report water-quality results. Selected student writings are also available via this home page. The Rivers Project home page contains a link to the e-mail address.

The Rivers Biology Unit

In *Rivers Biology,* students will learn about biological factors that indicate or are influenced by water quality and the quality of riverine habitats. They will acquire this knowledge not only through reading, class discussion, and written assignment, but in particular through laboratory work and field-trip observations involving their local river or stream.

In the laboratory, they will test river or stream water for such factors as fecal coliform—whose presence indicates external degradation of water quality—and oxygen—whose level significantly affects the ability of water-dwelling animals to survive in those waters.

They will need to take one or more field trips to a local waterway to collect and test the water for such factors, and to observe the biological diversity at the site. Central to the study of biological diversity of the stream will be the use of habitat assessment, biological indices, and diversity indices. Students will analyze their test results, determine water quality and biological diversity, and may share some of their information with other schools or state agencies using telecommunications.

BIOLOGY OF THE RIVER

The United States Environmental Protection Agency, the National Sanitation Foundation, and many individual state agencies have established parameters to measure the water quality of rivers and streams. Of those, the factors studied in *Rivers Biology* include many of the most commonly-used biological indices, physical measurements, and chemical tests, specifically:

- Benthic macroinvertebrate water tolerance index. A statistical evaluation of the relative predominance of organisms having varying levels of pollution tolerance.
- Diversity index. A statistical evaluation of organisms to determine the diversity of the riverine population.
- General habitat survey. Evaluation of stream or river quality based on observation of physical, cultural, and biological characteristics.
- Dissolved oxygen. The number of milligrams of oxygen gas dissolved in one liter of water. Oxygen levels determine which aquatic organisms can live in a body of water.
- Biochemical oxygen demand (BOD). The amount of dissolved oxygen consumed by organic matter or chemicals in a sample of water over a five-day period. The higher the BOD, the more quickly oxygen in the water is being depleted, reducing its availability for aquatic organisms.
- Fecal coliform. The number of fecal coliform bacteria present per 100 mL of water. The presence of fecal material indicates that the water may contain pathogenic organisms associated with human activity.
- Environmental assessment. The process of analyzing potential changes to riverine habitats, with recommendations to approve, alter, or prohibit those changes.

In order for students to grasp the biology of rivers, they must understand basic food relationships that occur within the aquatic habitat. Therefore, *Rivers Biology* covers aquatic food chains and food webs. Students must also become acquainted with, collect properly, and learn how to recognize riverine organisms. In this unit, common organisms are produced in graphic form for students, to help students practice recognizing specific organisms in significant groups, specifically benthic macroinvertebrates. Students learn, practice, and perform collecting techniques needed in order to sample adequately for benthic macroinvertebrates.

For each parameter, *Rivers Biology* gives students background information, supporting activities, and instructions on gathering and analyzing data. Teacher Notes for each lesson contain additional information and suggestions for the instructor. Informational activities support specific science-related skills, such as using statistics and conducting particular observations. Several types of assessment strategies support the lessons.

OUTCOMES FOR *RIVERS BIOLOGY*

In this unit, students will learn how to analyze river and stream water and habitats regarding (1) food chain and food web relationships; (2) benthic macroinvertebrate presence and water tolerance; (3) diversity of riverine populations; (4) general habitat information; (5) dissolved oxygen; (6) biochemical oxygen demand; (7) fecal coliform; and (8) protection of riverine environments through environmental assessment.

Using the activities, students can determine if their water is excellent, good, medium, bad, or very bad and can learn what interactions are taking place among organisms found in the riverine habitat. They can determine whether the biological diversity of their river is being impaired and whether human changes to the water or the watershed might impact water quality. In addition, students will see the wide variety of living organisms that make up the riverine habitat. Involvement with local river quality should help students become more environmentally aware and concerned about water resources.

ADVANCE PREPARATION

Before starting *Rivers Biology,* each teacher will need to gather numerous materials, become familiar with topics included in the unit, do preliminary planning for field trips, and assess student readiness for the project.

Student Preparation

This unit has been written for high school and middle school students in natural science and biology classes. Middle school students may require more instructional time per lesson than older students. Though students who have had algebra usually perform more comfortably with some parts of this unit, students without an algebra background have completed those parts of *Rivers Biology* successfully.

Teacher Preparation

Prior to introducing the *Rivers Biology* unit to your class, you should prepare by doing the following:

1. Read the entire unit.
2. Brush up on concepts used in this material.
3. Coordinate activities and curriculum with teachers considering doing other units of the Rivers curriculum.
4. Develop timeline for placing *Rivers Biology* in your year-long instructional schedule. Times noted for each activity are approximations; allow some flexibility.
5. Discuss the curriculum with your school administration and secure permission for the field trips and the acquisition of materials.
6. Find and visit an appropriate field site. Select a site with suitability for the activities, easy accessibility, and minimal safety hazards. If possible, select a site that may be impacted by some proposed change in the environment or human activities, or select a site that may face such consideration in the future. (For more guidance on this topic, see the Teacher Notes for Lesson 7.) When choosing your site, also remember that if you use a site near a state or local monitoring station on the water, your students can compare their data with data from the monitoring station. Obtain all necessary permissions from the landowner for students to visit the site.
7. As appropriate for your situation, send notes to parents informing them that students will be involved in river or stream field trips, as well as safe handling of river water. Ask for signature if appropriate.
8. Schedule field trips and prepare field-trip permission forms. Secure field-trip transportation. (See discussion of Field Trip Management later in this section.)
9. Collect all necessary materials and equipment.
10. Locate an established aquarium of plants and animals for students to observe, or set up an aquarium of living plants and animals and allow at least four weeks for it to become balanced before beginning Lesson 2. (For larger classes, consider having more than one aquarium.)
11. Make sure your laboratory is equipped with common safety devices, such as eyewash fountain, fire extinguisher, material safety data sheets (MSDS) for materials to be used, and first-aid kit.
12. Obtain maps of your site from the U.S. Geological Survey; also order the pamphlet *Topographic Map*

Symbols (one per two students). (For contact information, see Appendix C, Resources.) Other map sources include local or state agencies. For source information, contact your local Cooperative Extension Service. Global Positioning System (GPS) devices can be used to obtain latitude and longitude rather than maps. These are available through sporting supply stores or catalogs. Often, a local hunter, boater, or fish enthusiast will have one.

13. Enlist the aid of an expert (such as a conservation officer, planning commission member, Army Corps of Engineers employee, Fish and Wildlife representative, or other public employee) who works with river- and stream-related concerns and who is willing to address your students or lead them on a field trip.
14. Duplicate all materials to be distributed to students (including Student Information, Activity, and Assessment pages), punch holes for three-ring binders, and place in labeled files.
15. Several water-quality tests require special kits. Because of the increasing variety of test kits, this unit does not include specific instructions for kit use. Become familiar with operation of the test kits you will use; prepare student instructions for those kits. Laminating the instructions will protect them from spills in the laboratory and water in the field.
16. Select and obtain the special equipment needed for *Rivers Biology*. Special-equipment needs are discussed later in this section, and a list of suppliers is included in Appendix C, Resources.
17. Organize and update your collection of keys and identification books. The enclosed resource list offers a list of recommended identification books.
18. Generally, one waste container can be used for all the tests included in *Rivers Biology*. If your state or other jurisdiction requires separate containers for different types of chemical wastes at the levels produced in these lessons, prepare and label separate containers.
19. Get excited. Your enthusiasm will be contagious.

Field Trip Management

Field trips, particularly field trips around a body of water, require careful planning and management. Here are steps to make *Rivers Biology* field trips safe and successful. Adapt this list to fit your own circumstances and preferences.

Before the Field Trip

__ Visit and select an appropriate field-trip site.

__ Secure school administration permission for field trip.

__ Secure permission from landowner.

__ Contact expert for help and content assistance.

__ Publicize the event with local newspaper, television, and radio

__ Arrange for field-trip transportation.

__ Arrange for appropriate adult supervision on trip.

__ Prepare directions for field-trip drivers.

__ Develop an activity plan for the field-trip day.

__ Arrange for lunches or snacks, drinking water, and rest rooms.

__ Tell students what activities they will perform at the field site.

Distribute field-trip permission forms to students, and receive signed forms. (Permission forms should include request for information about student allergies and other medical needs.)

__ Give instruction on field-day clothing and personal items (journal, handouts, pencil, sunscreen, insect repellent, lunch, drinks, medications for individual allergies, and other needs).

Day Before the Field Trip

__ Make sure all student permissions have been returned.

__ Review activities that will take place during field trip, including extra activities, such as writing and litter pickup.

__ Each group assembles and prepares materials and equipment for their specific activities to be done at the river.

__ Pack all field-trip materials and equipment.

__ Assign students to complete reading of field-trip activity sheets.

Field Trip Day

__ Distribute directions to field-trip drivers.

__ Assign students to load equipment and materials for transportation to field site.

__ Assign students to clean and dry equipment and materials for return to the classroom.

__ Assign responsibilities to supervising adults.

__ Distribute safety and field equipment and materials.

__ Students do assigned testing and sampling activities.

__ Secure all water samples to be transported.

__ If students have completed assigned activities, have them help other groups, do riverside cleanup, or write in their journals.

After the Field Trip

__ Students complete lab work.

__ Wash and store all equipment.

__ Students complete calculations for their water-quality activity.

SPECIAL MATERIALS AND EQUIPMENT NEEDED

Each lesson and each assessment includes a list of the specific equipment and materials it requires. Appendix C lists suppliers of biological and chemical items approved by the Environmental Protection Agency (EPA). Here is some essential equipment not found in most school biology laboratories:

- Kick nets and dip nets for collecting aquatic organisms
- Collecting bottles, buckets, collecting pans, and insect forceps
- Plant and animal identification books
- Plant presses are optional, useful if you want the students to make a reference collection from the collecting site.
- EPA-approved test kits for dissolved oxygen. Ideally, each class should have at least three kits, but a procedure by which an entire class can use one set of test kits is described later in this introduction.
- Bacteria culture equipment for fecal coliform test. Most bacterial testing kits use a membrane filtration technique for counting bacterial colonies. Newly developing technology will surely change this process, making counting easier and quicker in the field. Most tests require incubating the bacteria for 24 hours in a hot-water bath held at a constant temperature.
- Safety goggles, aprons, gloves, eyewash, a first-aid kit available while testing at the field site, and waste containers for liquids and solids
- Alcohol-filled thermometers equipped with metal jackets
- Life jacket and tow line (rope) for each student if the waterway is not a small, shallow stream
- Waders or high rubber boots, for contaminated streams or when water is too cold to enter. Avoid hip waders, because they allow the student to enter water that may be too deep.
- Water-sampling device, purchased or constructed. One can be made by attaching at the end of a 3- to 4-meter pole a test-tube type clamp to which a glass or plastic bottle can be affixed. (More construction details are presented in *Rivers Chemistry*).
- A computer with modem and telephone hook-up so students can exchange data with other schools electronically and access the World Wide Web.

SUGGESTED SCHEDULE

The *Rivers Biology* unit is flexible: you can plan to do some of the lessons or all. Within a particular lesson, you can choose which of the activities suit your purposes. You can concentrate exclusively on *Rivers Biology* as a self-contained unit within your school year, or you can do a few lessons at a time, spread over a semester or more. Many teachers devote three weeks to *Rivers Biology* at the beginning of the year, continuing after that to spread the rest throughout the school year.

In Lesson 7, students will hold a mock public interest meeting. Review that material to decide what kind of audience, and, therefore, publicity you want for that meeting. If your choices dictate a long lead time for publicity before the meeting, do the first three steps of Introducing the Lesson in Lesson 4 Teacher Notes, then have students do the publicity for the meeting earlier, such as after completing Lesson 4.

River field trips are very important to *Rivers Biology*. The minimum number is one, but two trips, even short ones, really make your rivers study more valuable.

Students gain increased understanding of the factors that influence the health of the river environment. The more field trips, the better your students' experience will be. The number of times you sample the river and involve students in testing will depend on your school's curriculum and your commitment to the project.

In *Rivers Biology,* several lessons culminate in a field-site activity. Each field-site activity is preceded by one or more in-class information or activity sessions that can be combined or spread out over several periods as needed. You may complete several lessons up to the field trip activity, then do the field-trip-based activities for several lessons during one outing. Most field-site activities require less than one hour at the site. So, if the stream site is some distance away, or if only one or two field trips will be possible, you can complete all field-based activities in those one or two visits by having students carry out several field-site activities on one day. In planning field-trip time, take into account the time students will need in order to perform the assigned activities in the field; approximations are provided in the nearby table.

Field-Site Activity	Field-Site Test or Observations	Average Student Time Required in Field
Lesson 1	Optional field visit for observation (as assessment)	30–60 min
4.4	Determining the Benthic Macroinvertebrate Index for Your River or Stream	50–100 min
4.5	Completing a Riverine Habitat Survey	20–30 min
5.4	Measuring Dissolved Oxygen in a River or Stream	40–50 min
6.3	Measuring Fecal Coliform in a River or Stream	25–45 min

In *Rivers Chemistry,* students perform all nine tests of water quality used by the National Sanitation Foundation; based on those results they compute the overall water-quality index for that river or stream site. *Rivers Biology* includes those tests most directly related to the ability of the stream or river to support life—tests for dissolved oxygen, biochemical oxygen demand, and fecal coliform. Performing the additional tests covered in *Rivers Chemistry* (or obtaining data from another science class doing *Rivers Chemistry*) would allow students to compare the benthic macroinvertebrate water-tolerance results and the aquatic diversity survey in *Rivers Biology* with an overall water-quality index based on the complete set of water-quality tests.

USING ONE SET OF EQUIPMENT FOR AN ENTIRE CLASS

Ideally, during laboratory and field trip sessions each *Rivers Biology* class should be divided into at least four groups, and each group should have its own materials. To keep costs down, however, an entire class can use one set of test kits to conduct all tests on a single trip to the river or stream, using the steps following. (Teachers who use this method will need to make minor adaptations in testing and data-table procedures.) Over the course of the project, groups can be rotated so all students get experience doing all tests.

Day One: One Class Period of 40–50 Minutes

1. Divide the students into groups.
2. Demonstrate the operational procedure for each test.
3. Find location on maps.

Day Two: One Class Period of 40–50 Minutes

Students practice in lab becoming familiar with the equipment and the instructions for their individual topic or topics.

Day Three: Field Trip at the River or Stream

Group 1: Conducts the habitat survey and collects photographs and videos of the collecting activities.

Group 2: Conducts the benthic macroinvertebrate collecting along the ripple area using kick nets.

Group 3: Conducts the benthic macroinvertebrate collecting in pools and along edges of runs using the dip nets.

Group 4: Does the dissolved oxygen test at least three times and prepares at least three water samples for biochemical oxygen demand.

Group 5: Measures temperature and flow rate, each at least three times. Gathers other data such as precipitation, air temperature, and location of site on map.

Group 6: Prepares at least three water samples on which to do the fecal coliform test at the school lab.

After Return to the School Lab

Group 1: Records notes from habitat survey and prepares a form to archive. Prepare pictures and video information to archive or use as support documentation.

Group 2: Collects data from Group 3 and completes data table for benthic macroinvertebrate water tolerance index.

Group 3: Collects data from Group 2 and completes data table for diversity index.

Group 4: Places the collected river or stream water samples in a setting that will allow for the temperature-controlled five-day incubation period required for the biochemical oxygen demand analysis. Finishes data table for dissolved oxygen. Shares data with the groups preparing indices and the habitat survey.

Group 5: Organizes data and shares the information with the groups preparing indices and habitat survey.

Group 6: Prepares at least three samples for the 24-hour incubation period required for the fecal coliform test.

Day Four: One Class Period of 40–50 Minutes

Group 6: Finishes fecal coliform test. Shares data with the groups preparing indices and the habitat survey.

Each group goes through their inventory to determine whether all equipment is present and properly maintained.

Day Five: One Class Period of 40–50 Minutes

At the end of the five-day incubation period, Group 3 demonstrates how to determine biochemical oxygen demand.

SAFETY CONSIDERATIONS

While doing *Rivers Biology,* the teacher must take responsibility for and teach the students about safety in the laboratory and at the field site. Every state has governmental agencies responsible for the protection and monitoring of the environment. Persons involved in water quality should become familiar with their state and local environmental agencies. Information can generally be obtained by contacting the environmental agency at the state capital or, more often, at your regional office.

The U.S. Department of Labor's Occupational Safety and Health Administration (OSHA) is responsible for enforcing federal laws dealing with safety and hazards in the workplace. Specifically, federal regulation (29CFR1910.1200) requires "Hazard Communication Standards" to ensure that the hazards of all chemicals produced or imported are evaluated, and that information concerning their hazards is transmitted to teachers. This transmittal of information must be comprehensive and include container labeling and material safety data sheets (MSDS). In *Rivers Biology* (and *Rivers Chemistry*), therefore, it is assumed that when you receive laboratory chemicals you also receive the material safety data sheets.

Commercially available water test kits (see listing in Appendix C for HACH Company, LaMotte Company, and CHEMetrics, Inc.) are relatively safe for the students and the environment. Chemical waste should be disposed of properly, however. Even though the volume and concentrations of wastes are insignificant compared with a large river's volume, good environmental practices require that waste chemicals from these test kits NOT BE POURED OR DUMPED INTO THE RIVER. Specifically, the dissolved-oxygen test generally contains trace metals of toxic materials. The chemical waste from any tests should be stored in a container labeled "Waste." If the waste container is placed under an exhaust hood for several days, the water will evaporate and leave only a small volume of residue.

Outdoors collection of river or stream samples and running of chemical tests requires special safety considerations. Especially during the collection of benthic macroinvertebrates, attention must be taken that students not enter deep water. Your site may require that the person taking samples wear a life jacket and a tow line. If the water is likely to contain hazardous materials, students should wear protective gloves. They must wear safety goggles during chemical testing.

Know and discuss with students natural hazards in your field-study area. For instance, if poisonous reptiles such as rattlesnakes, copperheads, or water moccasins may be present, warn students to watch the ground and the surface of the water carefully for snakes.

If ticks may be present (more common in spring and summer in some areas), have students dress appropriately, avoid brushy areas as much as possible, and check afterwards. When approaching the stream, have students take caution to recognize and stay away from irritating plants, such as poison oak, ivy, and sumac. Look for and avoid stinging nettle, sometimes found along river and stream bottoms.

If possible include a litter clean-up activity at the end of each field day. This teaches students good stewardship and encourages good antilittering habits.

An example of a student guide on laboratory and field safety rules is included in Lesson 1. Students are asked to sign on the bottom of the page to verify that they have received and read the rules.

ASSESSMENT

Assessment is the opportunity for students to demonstrate what they have learned and to exhibit their ability to apply that knowledge in a meaningful way. The instructor has the responsibility to provide instruments that give students the opportunity to best show their knowledge and skills. *Rivers Biology* provides a wide array of types of assessment tools. Just as so many other aspects of this unit are flexible, so, too, are teachers welcome to select which assessment tools suit their preferences and circumstances.

Traditional assessment tools assess student knowledge of a subject in terms of how well the student has learned and can reproduce the facts provided in the lessons. In *Rivers Biology,* many of the short-answer questions at the end of Student Information Sheets and Student Activity Sheets are designed to meet these objectives. Teachers use these as in-class assignments and as homework. Sample answers for these sheets, which also include short writing assignments, are included in Appendix B. Teachers may use their own performance criteria and scoring rubric for such standard components of science classes as data handling, graphing, and lab performance.

Educators increasingly realize the importance of assessing the extent to which the student can apply the instructional information to problem-solving situations; in other words, teachers want to know whether their students can really use what they've learned. Alternative assessment, often referred to as authentic assessment, attempts to determine if the student's knowledge is "usable." Instead of multiple-choice or other paper-and-pencil tests, the Rivers Project curriculum emphasizes authentic assessments tools. Authentic assessment assignments usually do not have a single answer, and students must employ critical thinking skills and creativity. Two of the more popular forms of authentic assessment are performance assessments and portfolio assessments. In both, students demonstrate how they use the classroom instruction to find a logical solution to a problem.

Performance Assessment

Performance assessment requires students to actively demonstrate their ability to apply the information presented in the curriculum. Individual students or groups of students work cooperatively to demonstrate the extent of their understanding. Appendix A includes Teacher Notes and Student Information Sheets for assessment tools that apply throughout the unit (such as collages and field-study skills) or as unit-end assessments (such as online river quality research and preparing data for online submission). Several lessons in this unit include lab-based or research-oriented assessments. Such assessments include performance criteria for students. The Teacher Notes for each corresponding lesson contain a suggested scoring rubric for that assessment.

Portfolio Assessment

An academic portfolio is not an accumulation of all the assignments and tests of a grading period; rather, it is a collection of selected documents organized as evidence of a student's cumulative learning. Like an artist's or model's portfolio, the academic portfolio is an opportunity for the students to demonstrate the depth and breadth of their learning. The student takes responsibility for selecting what to include in the portfolio and how to organize it. Students must feel a sense of ownership for their portfolio. The goal of the portfolio process is that the student reflects on the personal learning process and recognizes learning as a process that extends beyond textbooks and lectures. Teacher Notes and Student Information on portfolios are included in Appendix A.

TEACHER NOTES

LESSON 1 How to Begin a River Study

Focus

Students will learn how to use a journal in a scientific investigation, safety guidelines and techniques for laboratory and field work, and about the water resources in local and worldwide environments.

Learner Outcomes

Students will:

1. Learn how to keep a journal for personal and class work.
2. Learn safety precautions for lab and field work.
3. Become familiar with water resources locally and on Earth.

Time

Three class periods of 40–50 minutes per period

DAY 1: Student Information 1.1: Introduction to Rivers Biology
Student Information 1.2: Journal Writing
Student Information 1.3: Safety Guidelines and Contract

DAY 2: Student Information 1.4: The Facts About Water

DAY 3: Student Activity 1.5: Water in Your Area (Allow additional time for speaker and research as appropriate.)

Advance Preparation

Prepare to supply students with Student Information and Activity sheets 1.1 through 1.5. If desired, prepare two copies of Student Information 1.3 per student, so that each student may keep one copy and return one signed copy. If appropriate, add parental signature line.

Gather all necessary materials for this lesson. Make arrangements for each student to have a journal. If possible, have sample journals available for students to view.

Decide whether to include project-long assessment tools presented in Appendix A, such as portfolios or collages. If so, prepare to supply students with Student Information sheets about the selected assessments. If desired, prepare a copy of the Glossary for *Rivers Biology* for each student to reference throughout the unit.

Collecting information on state and local water resources is absolutely necessary. Contact your county soil and water conservation office or the state department of natural resources, environmental protection agency, or department of conservation for materials on water resources. Try to obtain classroom sets of such materials for students to use for research and completing classroom activities.

Arrange for an expert from the state, county, or city to talk about water resources. Ask this speaker to include such topics as precipitation patterns, water quality in local rivers and streams, groundwater and surface water, flooding and erosion in the local area, and local water sources. If you are not able to obtain a speaker, prepare a lecture that will cover these topics and others pertinent to Student Activity 1.5: Water in Your Area. Add questions to Student Activity 1.5 or make revisions to this sheet as appropriate to reflect available information.

Gather resources for students to use in determining the local climate. You can do this by using an almanac, local weather service, or appropriate World Wide Web sites, such as the National Climatic Data Center (see Appendix C). (Either retrieve the data yourself or have students collect the information, as appropriate.)

Safety and Waste Disposal

If students will go to the field site as part of the assessment for this lesson, make sure they are familiar with the safety procedures covered in Student Information 1.3 beforehand.

Materials

Student Information 1.2: Journal Writing
Per student
journal with bound pages

Student Activity 1.5: Water in Your Area
Per group
state map showing rivers, streams, and lakes
compass
Per class
information on local water resources
information on state water resources
information on local climate
reference books such as an encyclopedia, almanac, atlas, and gazetteer

Vocabulary

aquatic biologist
aquatic habitat
condensation
environment
evaporation
fauna
field biology
flora
fresh water
groundwater
habitat
hydrologic cycle
infiltration
marine biologist
material safety data sheet (MSDS sheet)
organism
potable water
precipitation
river biologist
runoff
saline
sensory perception
sewage
surface water
topographic map
transpiration
water cycle
watershed
wetlands map

Background for the Teacher

For Student Information 1.3: Safety Guidelines and Contract, you may need to alter the safety guidelines for lab and field work, and the student contract, to fit your particular situation.

Introducing the Lesson

1. Have the students read, discuss, and answer the questions for Student Information 1.1: Introduction to Rivers Biology. (Answers for student sheets are in Appendix B.)
2. Have the students recall information that happened at school, in the newspaper, or perhaps the weather one week ago. Emphasize that, in science, keeping a journal is important so investigators can recall past information readily.
3. Make sure each student has a journal. Have students read, discuss, answer questions, and prepare their journals as described in Student Information 1.2: Journal Writing. For specific teaching activities on journal writing, see *Rivers Language Arts*. For specific teaching activities on differentiating between observations and inferences when making journal entries, see *Rivers Earth Science*.

Developing the Lesson

1. Ask students if they have ever heard of someone being injured at work or home while performing a specific task. Have them discuss what could have prevented the accident.

2. Have students read, discuss, and answer questions on Student Information 1.3: Safety Guidelines and Contract. (This material was originally developed for *Rivers Chemistry*.) Show students the location and use of safety devices. Emphasize the necessity for safety cautions in laboratory and in the field. Have students sign and return the safety contract. For additional specific information on laboratory and field skills, data accuracy, and data use, see *Rivers Chemistry*.
3. Have students read, discuss, and answer the questions for Student Information 1.4: The Facts About Water. (Portions of this handout were developed for *Rivers Earth Science*.) Discuss how much students know about water resources on Earth. For additional specific teaching activities on the hydrologic cycle, and on the limited amount of fresh water on Earth, see *Rivers Earth Science* and *Rivers Geography*.
4. If desired, distribute a copy of the *Rivers Biology* Glossary to each student. You may have students look up definitions of new terms. Discuss the vocabulary with the class as appropriate.
5. Have a local water environmental specialist or other expert speak to the class about state and local water resources. If a local expert is not available, prepare and deliver a lecture on these topics. Have the students take notes in their journals during all presentations. If desired, you may encourage students to ask questions, even having them prepare questions in advance. For specific teaching activities on interviewing, see *Rivers Language Arts*.
6. Have students complete Student Activity 1.5: Water in Your Area. Encourage them to base their responses on the presentation and on other resource materials available in the classroom (as well as outside). Allow additional research time as appropriate. If desired, have students work in teams, each on a separate section of the Observations.
7. Discuss the results of Student Activity 1.5. Emphasize the importance of monitoring the quality of waterways and recording good data or observations. Relate this to biology by noting how water quality impacts living things. Refer to how specific human activities affect animals and plants living in or near the water.

Concluding the Lesson

1. Have students note in their journals their predictions for the water quality of the river or stream they will be testing. Do they have any concerns about the water they observed during the field-site visit? Tell them that they will be referring to these predictions again after they have completed the river study.

Assessing the Lesson

1. If possible, have students do a short field-site visit at some local river, stream, or lake, or some open area adjacent to the school. Have students practice making observations. Remind them to follow all safety precautions. As homework, have them write a short paper connecting their observations with what they have learned about water resources so far

in *Rivers Biology*. Review their journal entries for organization and completeness. Have students share with the class one or two journal entries they feel are important. Here is a suggested scoring rubric.

Score	Expectations
0	Observations missing from the journal; no questions developed; no paper written.
1	Observations made in the journal. Journal page set up but not in an organized way. Paper partially developed, with little relationship between observations and information learned.
2	Observations made in the journal. Observation procedure followed in an organized way. Paper shows some relationship between observations and factual water information.
3	Generally satisfactory observations made in the journal. Observation procedure followed in an organized way. Paper makes meaningful connections between observations and factual water information.
4	Excellent observations made in the journal. Observation procedure followed in an organized way and in depth. Paper makes understandable connections between observations and factual water information. The writing completely develops a philosophy about water study and good stewardship.

2. Distribute and discuss with students information on any unit-long assessment tools you plan to use, such as portfolios or collages. (You may distribute the detailed handout about portfolios in either Lesson 1 or Lesson 7.) For Teacher Notes and sample students sheets on such assessment tools, see Appendix A.
3. If students will be doing river or stream collages, have them start collecting materials.

Extending the Lesson

1. Arrange a field trip to an industrial laboratory or invite a laboratory technician to visit the class to discuss safety and laboratory techniques.
2. Arrange a field trip to a water-testing lab or water-treatment facility, or invite a water-quality technician to visit the class to discuss the performance and reporting of water-quality data.
3. Have students start a newsletter or bulletin board that informs others in their school about the local river or stream, periodically updating information as the project progresses. For specific teaching activities about writing newspaper articles, see *Rivers Language Arts*.
4. Have students write in their journal a short story, poem, or paragraph that summarizes or characterizes some important aspect of Lesson 1. For specific teaching activities about poetry writing, see *Rivers Language Arts*.

Name

Introduction to Rivers Biology

When you visit your local river or stream for a day of fun, you've probably rejoiced when you've noticed wildlife in or by the water. Perhaps you've seen eagles, trout, turtles, or little unidentified bugs skimming the surface. What you may not know is that many species live in those waters and that their presence indicates the health status of that river or stream.

When people study the chemistry of a river or stream, they collect samples of the water to test for the quality of the water in that river environment. (In this context, **environment** means all the conditions and circumstances surrounding and affecting an organism or group of organisms.) The results reflect the conditions of the water just at the time of collection; they're just a snapshot. Pollution may have entered the water upstream and not yet reached the site being tested. The only way to be sure of having the complete picture is to monitor the water on an ongoing basis.

When people study the biology of a river or stream, they evaluate the organisms living in that environment. The existence of small river organisms depends on the quality of water flowing by them; thus they provide a panoramic view of the water quality. Using organisms as indicators of a stream's health reveals the total environment of that stream, not just a snapshot of a particular moment. A **river biologist** studies the organisms of rivers, ponds, lakes, and the surrounding wetlands.

Biology and the River

The river or stream you are about to study is home for many plants and animals. Different rivers and streams host different organisms; even different spots at the same site host different species. Some plants live only in wetlands or along the edges of a river, because they have developed structures adapted to abundant water sources. Willows and cottonwoods, for instance, tend to grow with their roots in water. Water lilies and duckweed can survive only in water. Pollution has a great impact on only a few river-environment plants; even then, that impact must occur over a long period of time or be very toxic. What usually impacts plants more is a change in the **habitat,** the area or type of environment in which that organism or ecological community lives. For plants, such change might be the draining of a swamp or building of a levee.

Water quality does not greatly affect air-breathing animals of the water such as birds, turtles, snakes, beaver, muskrat, and otter. Such organisms are mobile, so if pollution disturbs one area, they'll move to a less-affected site. Small invertebrates, fish, and mussels living submerged in the water, however, cannot move to another part of the river or stream. So what flows into the river has a great impact on them. For example, if a clean and rocky tributary stream receives a load of silt and soil, the fish and invertebrates that live there may die of suffocation. Thus, organisms that live in the water serve as indicators of the quality of the water they inhabit.

Human activity has changed, and continues to change, North American rivers. Becoming familiar with river environments and the biological interactions in these settings will enable you to participate in decisions about the health and future of such sites. With this biological knowledge, you will learn techniques and analysis for measuring rate of improvement or degradation of river and stream environments. By becoming more aware of the ways of the river and stream, you will learn not only the importance of protection and nurturing such waters, but the means of doing so.

About *Rivers Biology*

The first six lessons of *Rivers Biology* deal with information and procedures focused on river organisms and habitats. You will study the interrelationships among such organisms. You will learn how to detect, collect, categorize, and enumerate varieties of invertebrates. You will perform surveys at a local river or stream site and interpret the quantity and diversity of such organisms as an indication of water quality. You will assess the suitability of that habitat at that site for river-oriented organisms. This will include looking for signs about whether human changes to the water or the watershed might be impacting water quality or biological diversity. You will also study and perform several water-quality tests that relate directly to the suitability for river organisms.

In Lesson 7, you will use this accumulated knowledge to take a broader analytical look at a specific site on a local river or stream. In considering whether to permit a proposed (real or hypothetical) human alteration to the environment, you will assess its potential impact on the river environment and its inhabitants.

Rivers Biology involves more than traditional classroom and laboratory work. You will be working with your stream directly, probably getting wet and dirty. Using the Internet, you may share some of your findings with other students and scientists across the state and country, forming a database for succeeding classes. You will be not just studying for a grade, or even to gain information or skills, but to contribute to the knowledge of those who will be making decisions about your rivers and streams.

Questions

1. Write a short paragraph comparing the biologist's and the chemist's view of the river.
2. Describe what you hope to learn from this study of *Rivers Biology*.
3. Describe the nearest stream or river. Mention all the plants, animals, and other living beings you believe live in that environment.

Journal Writing

An **aquatic habitat** is a water-oriented setting that contains a variety of **flora** (plants) and **fauna** (animals), some seen and some unseen. Using all your senses to investigate this habitat can lead you to a more comprehensive awareness of what lives in wet places and of the relationships among the organisms (living entities) in this environment. **Sensory perception** (seeing, hearing, smelling, feeling) of an aquatic environment can be the beginning of a better understanding of that area and of rivers. Making good observations is the key to becoming a good biologist.

Using a journal to record your observations begins the process of river awareness. Can you recall exactly what you did or observed several days ago? Probably not; human memory is often faulty or incomplete. So, keeping a written journal in which you record your thoughts and activities about your river study is important. You will start *Rivers Biology* by organizing a journal for your own use.

Field Biology and Journal Writing

Field biology, the study of living things outside or in the field, involves a process that is time-consuming, expensive, and frustrating. To make this process more productive and efficient, biologists record their field observations as accurately and with as much detail as possible.

The essence of the field journal is the data the researcher collects; such data should reflect the specific variables being studied. In the journal, the writer records events as they happen in the field. A journal provides the ability to use the recorded information at some future date. What separates journal producers from diary writers is the systematic recording, categorizing, and retrieving of information that goes on before, during, and after each field event.

Keeping a Journal

The materials in the journal of a field biologist will appear different from that of an anthropologist, sociologist, or chemist. However, similarities in structure appear among all scientific journals. The journal should be bound so that pages cannot be removed or inserted. The front two pages should be a table of contents, filled in as the journal proceeds. Here's how to organize your journal:

1. Write your name in the front, along with "Rivers Biology."
2. Number each page consecutively.
3. Label pages 1 and 2 as "Table of Contents."
4. Beginning on page 3, and at the beginning of each new journal entry, place the following headings:

   Date:
   Time:
   Temperature:
   Wind direction and velocity:
   Weather:
   Location:
   Purpose of the observation: (leave several lines)

Descriptor key:	Sight(Si)	Hearing(H)	Smell(Sm)	Touch(T)
	(Color)	(Color)	(Color)	(Color)

   Data:

Depending on the field situation and purpose of the observation, you may not be able to make entries for all the items in the preceding list, such as wind velocity, but record as much data as possible. Unless you have quantitative data, use a general descriptor (such as "slight wind from Northeast") instead of an exact number.

To reduce the amount of writing required, you may use a descriptor key, based on either abbreviations or color coding. For instance, you may reduce variables to a set of abbreviations or code. As shown in the preceding list, sight becomes Si, hearing becomes H, and so forth. Some researchers use color-coded highlighting pens to classify information by color, making visual retrieval quicker. If you do this, note your color code at the beginning of your journal entry, as indicated in the preceding list by the word *color*.

Write all journal entries legibly in ink. Do not erase or completely obliterate any entry. If you want to make a change or revision, draw a line through an unwanted entry and write in your correction. Add the date and your initials to indicate you are the person making the correction.

As with any new skill, to do journal writing well, you must practice it. Bring your journal to every class meeting, and take it into the field as well. Your journal will be your own record of your progress during this course of study. Make an entry in your journal for each day class meets (more times if desired), reflecting on each day's activities, and relating what you have learned to any information you have already acquired. At the end of each activity, you will find suggestions for journal entries.

Include in your journal any equations, local examples, and other information discussed in class that is not part of the student information or student activity sheets. After completing an activity, write a summary that includes analyses of experimental data and your conclusions. If you are in the field, record the weather conditions, wind velocity (mild, gusty, and so forth), approximate

DRIFTWOOD

"We can never have enough of nature."

Henry David Thoreau

water depth, plants and animals at the site, and any other observations you feel are significant.

Observation entries are straightforward, containing no broader context or analysis. Include more than just data or scientific observations in your journal, however. In a special section or at the end of each day's writings, also record your personal impressions or reactions; summaries or critiques of class activities; assigned or independent study topics; and anecdotal information from your class work or fieldwork. Write about special happenings, triumphs, and projects that didn't work out as you planned or expected.

The purpose of your journal is to help you see the broader meaning of your river study and reflect on its importance. It is your personal repository for information that you may want to retrieve in the future.

DRIFTWOOD

"For in the end, we will conserve only what we love, we will love only what we understand, we will understand only what we are taught."

Baba Dioum
Central African conservationist

Questions

1. List as many song titles as you can that involve water.
2. List as many book and poem titles as you can that involve water.
3. Does anyone in your family or circle of friends keep a diary or journal? If so, talk to them about how they record information in the diary.
4. Clarence Olson has been keeping a journal of the weather in Richey, Montana, for more than 30 years. He does not report this information to anyone. He just keeps the data and has saved the diaries. Are they of any value? How might they be?

STUDENT INFORMATION 1.3

Name

Safety Guidelines and Contract

Laboratory work and fieldwork require specific, and different, kinds of safety precautions. By following the instructions in this material, you will be able to work safely in either type of setting. Once you have reviewed these materials, signify your agreement to follow these safety guidelines by signing your name to the Safety Contract.

Laboratory Safety Guidelines

1. Never perform an experiment in the laboratory without the permission and supervision of your teacher.
2. Never eat food or drink beverages in the laboratory or while collecting on the river.
3. Before performing any laboratory procedure, read the instructions and review the safety precautions.
4. Always wear safety goggles in the laboratory or when doing field chemical testing. Ordinary prescription glasses do not provide adequate protection from fumes or splashed chemicals.
5. Never allow chemicals or any hazardous material to come in contact with your skin. Wear protective gloves when handling chemicals or water that may be contaminated.
6. Shoes, not sandals, should be worn in the laboratory to protect your feet in the event of a spill or breakage. Waders or boots should be worn in the stream. Never enter the stream with bare feet.
7. If working with an open flame, tie back long hair and loose clothing. Make sure no flammable materials or vapors are nearby. Use tongs or insulated gloves to pick up equipment that has been heated recently.
8. Know the location of material safety data sheets (MSDS), portable eye-wash fountain, fire extinguisher, and first-aid kit and how to use each properly. **MSDS sheets** describe the proper safety precautions to observe when using specific chemicals.
9. Use a spatula for transferring solids, and use appropriate glassware for transferring liquids.
10. Never use dirty, cracked, or chipped glassware.
11. Never pick up broken glassware with your bare hands. Sweep broken glass into a dustpan and place in a waste container provided by your teacher.
12. If your skin comes in contact with acid or another chemical, immediately wash with soap and rinse the area thoroughly with water.

13. If acid spills in your work area, inform your teacher immediately.
14. Dispose of excess chemicals or wastes as directed by your teacher. A waste container should always be available either in the field or in the laboratory. **Never dispose of wastes into the water.**
15. Never return excess chemicals or reagents to their original containers. Clean up packaging materials from the test kits by placing them in the waste container.
16. At the end of each lab or field period, clean up your lab area and the equipment. Then wash your hands with soap and water thoroughly.

Field Safety Guidelines

1. Stay within the area directed by your teacher. Do not cross private property without the teacher's permission.
2. Always stay with your group. Conduct all sampling and analyses under the supervision of a teacher or your teacher's authorized aide.
3. Never engage in horseplay or distracting activities.
4. Wear safety goggles and protective gloves when conducting chemical or biological tests. Wear gloves and boots when coming in contact with water of unknown origin. Wear insect repellent and sunscreen if the conditions warrant.
5. Make sure you know the location of the first-aid kit and portable eyewash station. If you have any outdoor allergies, bring along any special medication that may be necessary in case of an allergy attack.
6. If working on a waterway bigger than a small, shallow stream, wear a life jacket and tow line.
7. Avoid walking or jumping on unstable or slippery ground, rocks, or fallen trees. Walk on the path when one is available.
8. Never drink from or wash food in a waterway. Bring water from the lab for washing and drinking.
9. Place all wastes in containers provided by your teacher and take the wastes back to your school for proper disposal. Do not leave waste at the field site; do not dump waste into a waterway.
10. After you have completed your fieldwork, wash your hands with soap and **potable** (POH tuh bul) **water** (water suitable for drinking).

Safety Contract

I have read and understand all the laboratory and field safety rules, and I agree to follow these rules. If I fail to follow these rules, I will not be allowed to do the activities.

____

Student's Signature

Questions

Read the following paragraph.

Sandra finished her fecal coliform lab. She put away her safety goggles and was cleaning up and washing down the area. The Petri dish for the tests were ready to be placed in the incubator. She wondered what would happen if acid were placed in the dish. She opened the Petri dish with her hands and poured in a small amount of acid that was nearby. It turned a dark color and began to smell. She quickly dumped everything into the sink, hoping her teacher had not noticed. She looked at her wristwatch; only a minute until bell time. She knew if she took time to clean her lab area and wash herself up, she wouldn't have much time to talk with Rashad in the hallway, so as soon as the bell rang, she walked right out the door.

1. Make a list of lab safety rules that Sandra did not follow.
2. Describe at least one negative impact that could have resulted from each broken safety precaution.

Read the following paragraph.

Jack thought it was great being outside instead of sitting in class. As he looked around a fallen tree along the river, he noticed a glittering metal object. "I think that's a fishing lure," he said to himself. Jack slowed down and waited until his teacher and classmates were ahead of him at the collecting site. He quickly left the path, climbed the log over the water, and pulled up the object that had attracted him. Sure enough, it was a brand new fishing lure. A great find! He crawled back along the log and onto the path. Running to catch up, he rejoined his classmates just as they got to the test site, thinking, "Today's my lucky day."

3. Make a list of field safety cautions Jack ignored.
4. What could have happened to make it Jack's unlucky day?

STUDENT INFORMATION 1.4

Name

The Facts About Water

Ever since water was formed during early days of the Earth's development, its volume has remained essentially unchanged. Nearly 75 percent of the Earth is covered with water, but it is distributed unevenly. About 1359 million cubic kilometers (326 million cubic miles) of water are spread around the planet.

Forms of Water

Over 97 percent of the water on Earth is in the oceans, an estimated 1321 million cubic kilometers (317 million cubic miles). Because this water is salt water, it cannot support most human uses, land-dwelling species, or freshwater organisms. (These species and activities need **fresh water,** water with a low salt content.)

Ice caps and glaciers make up 2.15 percent, of the total water supply, 29.2 million cubic kilometers (7 million cubic miles). Today, 77 percent of all the water not in the oceans is in glaciers and ice caps. If the continental glacier of Antarctica were to melt, the level of the oceans would rise by 60 meters, or almost 200 feet.

Surface water includes all fresh water on the surface of the Earth, such as lakes, rivers, and streams. It makes up only approximately 0.017 percent (about 230,493 cubic kilometers or 55,300 cubic miles) of the Earth's water. Of this, running water in rivers and streams comprises only about 1250 cubic kilometers (300 cubic miles) of water, or 0.0001% of the total. Freshwater lakes contain about 100 times this amount (125,000 cubic kilometers or 30,000 cubic miles; 0.009 percent).

The remainder of the surface water—104,200 cubic kilometers (25,000 cubic miles, or 0.008 percent)—is in **saline** (salt-containing) lakes and inland seas. Their salt content is too high to permit use for drinking or irrigation.

Groundwater, also called subsurface water, is any water that collects beneath the Earth's surface, generally within 4 km (2.5) miles of the surface; it is replenished by surface water seeping into the Earth. It comprises 0.625 percent of the Earth's total water. Scientists generally estimate that slightly more than 8 million cubic kilometers (2 million cubic miles) of water is stored in cracks and pore spaces in the soil and rock.

The very small portion of the Earth's water that makes up rivers, streams, and lakes is a thin thread that supports most land life. Humans also draw water from groundwater sources for drinking water and agriculture. (Agriculture

DRIFTWOOD

DO YOU KNOW?

What percent of a living tree is water?

What percent of your brain is water?

How many liters of water are needed to produce one egg?

How many liters of water does it take to process a meal of a quarter-pound hamburger, an order of fries, and a soft drink?

How many liters of water can 1 liter of gasoline contaminate?

(Answers on page 15)

consumes over 40 percent of North America's water.) The vast amount of water used by humans (78 percent), however, comes from surface water.

Water Is Life

Scientists typically study oceans separately from freshwater communities and environments, because the two systems have such different chemical compositions and natural forces acting upon them. Generally speaking, **aquatic biologists** study freshwater organisms and their environments. **Marine biologists** study saltwater organisms and their environments.

For both marine and aquatic organisms, water equals life. A study of any habitat, such as the desert, tundra, prairie, forest, wetlands, or tidepools, shows that availability of water is the key ingredient in providing for an abundance and variety of living organisms.

Where water is found in your area you can also find an abundance of life forms. Taking a visit to that river, stream, or pond, and learning about the organisms living there, will allow you to experience water's special importance to life and its diversity.

Driftwood

NOW YOU KNOW

Seventy-five percent of a living tree is water.

Seventy-five percent of your brain is water.

Producing an egg requires 480 liters (120 gallons) of water.

That fast-food meal requires 5600 liters (1400 gallons) of water.

Just 1 liter of gas can contaminate over 3 million liters of water.

Water Gets Around

The water on Earth is constantly changing from one state to another. When warmed, ice and snow melt into liquid. If the temperature increases enough, water changes into water vapor, which rises into the lower atmosphere. When cooled, water vapor in the atmosphere condenses into **precipitation** (prih sihp uh TAY shuhn)—rain, sleet, or snow that falls to the surface of the earth. The water may drain or flow off the land surface, forming **runoff.** Some of this water may seep into the soil, a process called **infiltration.** When the sun shines, the soil, and water, absorb its energy again.

As that water absorbs the sun's energy, it vaporizes, by the process of **evaporation** (ih vap uh RAY shuhn). Thus, the cycle starts all over again. As the water vapor cools, it releases energy and changes back into a liquid state, by the process of **condensation** (kahn dehn SAY shuhn). The endless cycling or interchange of water among the oceans, the lower atmosphere, the land surface, and underwater reserves is called the **water cycle,** or **hydrologic** (hi druh LAHJ ihk) **cycle.** The process in which water evaporates from plants is called **transpiration** (trans puh RAY shuhn).

This hydrologic cycle holds great importance for rivers and their species. When snow melts or rains come, then the rivers are full and life is good for the organisms that depend on or live in water. The flooding itself is important for some species, because the rising waters spread or nurture them.

Hydrologic events may significantly affect the water quality of the river or stream. For instance, low water flows magnify human impact: drought concentrates bad chemicals and sewage outflow into the waterway. (**Sewage** is human waste material carried in water through sewer pipe systems.) In times of low water, human water usage may leave many species and natural communities without enough to thrive, or even survive. In times of excess rain, runoff and rising waters may carry many pollutants into the stream, ranging from untreated sewage to drowned animals.

Questions

1. How much fresh water is available for use by humans?
2. How do you think fresh water is replaced on the land?
3. What is the significance to humans of realizing how small a proportion of the Earth's water is available as fresh water?
4. Using Figure 1-1, match the appropriate label with the letter that represents that aspect of the water cycle or related concepts.

condensation	infiltration	surface water
evaporation	precipitation	transpiration
groundwater	runoff	ocean

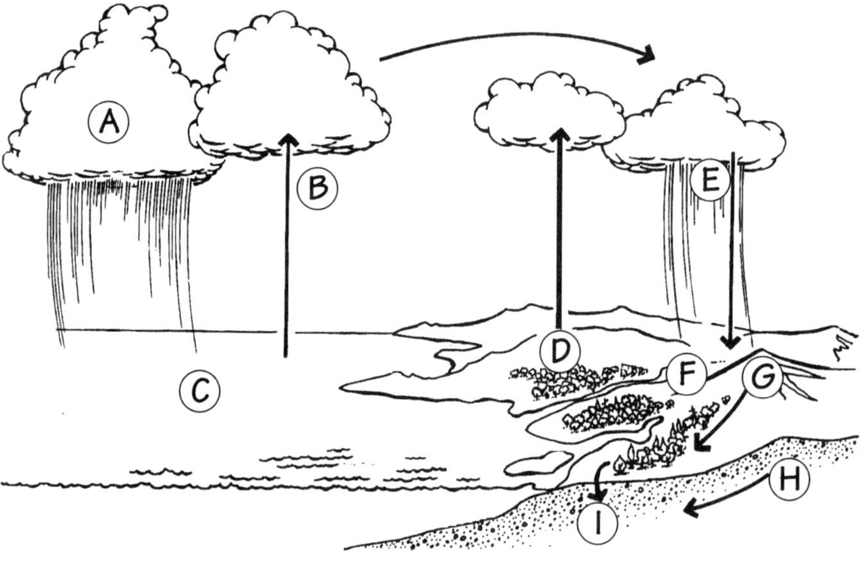

Figure 1-1: Hydrologic Cycle

5. Using Figure 1-1 as appropriate, name four places in the water cycle where local pollution could be introduced that may adversely affect a local river or stream.

Name

Water in Your Area

Purpose

To gather and evaluate water- and river-related information pertinent to your state and to your local area.

Background

To understand how and why your river or stream flows as it does, you need to understand not only the water cycle in general but also the water resources available to the **watershed** surrounding that specific river or stream. (The watershed is the area drained by a river system; the land area from which water flows toward a common stream in a natural basin.)

The geological, hydrologic, and seasonal changes that affect the river influence how both natural and human activities impact that water. You can find information about local and state water resources from libraries, state offices and Internet sites, and resource professionals. The United States Geological Survey (USGS) and other public agencies can provide topographic maps and wetlands maps for your area. **Topographic** (tahp uh GRAF ik) **maps** are large-scale, to show detailed natural and cultural surface features. **Wetlands maps** are special overlays for topographic maps that delineate and identify types of wetland areas. Putting together pieces of this river puzzle will allow you to appreciate its influence on you and your city or town.

Procedure

1. Using the materials provided by the teacher and collected from the suggested resource. persons and sites, answer the questions in the Observations section about state and area water resources, precipitation, and water quality.
2. Place the answers to the questions in your journal, or in a report as directed by your teacher. Use complete sentences.
3. Make notes on any other pertinent information you believe might have interest or value at a later date.

Materials

Per group

- state map showing rivers, streams, and lakes
- compass

Per class

- information on local water resources
- information on state water resources
- information on local climate
- reference books such as an encyclopedia, almanac, atlas, and gazetteer

Observations

Location of Water Resources

1. The major rivers in the state are ____________________. How long in kilometers is each?
2. The biggest river in the state is the ______________.
3. What rivers in this area are within a two-hour drive of school? _________ .

4. The largest lakes in the state are ________________.
5. The largest lake near the school is ________________.
6. _______ kilometers of streams are found in the state. Name streams that run near the school.

Precipitation

7. This part of the state is likely to have an annual precipitation of approximately _______ millimeters (mm) but has had as much rain as ________ mm and as little as ________ mm in a year.
8. During what season does most of the rain fall?

Water Quality

9. The two most common problems for the water quality of lakes and rivers in your state or area are: ________________ and ________________.
10. In most states, some government department ranks the streams and rivers in the state, such as by fish advisory or by water quality. If possible, review your state's official map or chart of water-quality ratings for rivers and streams. List the water-quality rating for each river and stream in your local area.
11. What is the source for drinking water for the school? your home? the town?
12. Where does your sewage go from each of these places?

Analysis and Conclusions

1. Explain why you think you live in a water-rich or water-poor state or area.
2. What is the relationship between the directions that the rivers and streams flow in your state (or local area) and the topography of the land?

Critical Thinking Questions

1. How do the climate or precipitation patterns in your area influence the health of your local river or stream?
2. If the weather were to change so your area received more rain annually, how would that impact your city and school?
3. How would a drought impact your area? Be sure to consider potential changes in water usage, other human activities, animal and plant communities, and the area's visual appearance.

Keeping Your Journal

1. In terms of water resources, in which part of the state would you most like to live? Why?
2. Which information was the most difficult to obtain for this report? Can something be done to make that job easier?
3. What was the most significant fact you found during this search?

LESSON 2 River and Stream Ecology

Focus

Students will discover that the health of a river or stream ecosystem depends on interactive food chains and webs. They will begin to study the interdependence of plants, animals, and their nonliving environment in an ecosystem. Through a series of activities that model the real aquatic world, students will begin to experience the complexity of energy as it flows through the environment.

Learner Outcomes

Students will:

1. be able to construct food chains or food webs and describe interactions taking place.
2. observe and describe interactions taking place in an aquatic ecosystem.
3. be able to describe examples of variety and diversity among organisms from aquatic environments.

Time

Five class periods of 40–50 minutes per period

DAY 1:	Student Information 2.1: River and Stream Ecology
DAY 2:	Student Activity 2.2: How Does Your Aquarium Rate?
DAY 3:	Student Activity 2.3: How Does a Food Web Work? Student Activity 2.4: Food-Chain Rummy
DAY 4:	Student Assessment 2:5: They All Hang Together
DAY 5:	Student Assessment 2:5: They All Hang Together

Advance Preparation

Prepare to supply students with Student Information, Activity, and Assessment sheets 2.1 through 2.5. Gather all necessary materials for this lesson. Prepare overhead transparencies of Teacher Figures 2-1 and 2-2, located at the end of these Teacher Notes.

For Student Activity 2.2: How Does Your Aquarium Rate?, review the established, balanced aquarium located or prepared before beginning *Rivers Biology*. Try to identify specific food chains and food webs within that aquarium ecosystem.

For Student Activity 2.4: Food-Chain Rummy, for each group of students, you will need three photocopies of each of the organism picture cards on pages 25 and 26 at the end of these Teacher Notes. (For instance, if you have 6 groups of 4 students each, you would need 18 photocopies of each figure.) For each group, you will also need 3 photocopies of the sun card on page 26 at the end of these Teacher Notes. Laminate and cut out the cards. For each group, place three sets of cards in a resealable bag. (You will use one set of these cards for Student Activity 2.3 also.)

Materials

Student Activity 2.2: How Does Your Aquarium Rate?

Per two students
medicine dropper
depression slides
microscope
Per group or class
balanced aquarium, open, (assembled 4 weeks earlier), containing:
live fish
live plants
live snails
aquarium gravel
other appropriate organisms
identification books showing microscopic aquatic organisms

Student Activity 2.3: How Does a Food Web Work?

Per student
colored pencils or markers
Per class
set of 15 organism cards (in resealable plastic bag), made from images on pages 25 and 26
15 safety pins (or pieces of masking tape)
20–25 pieces of string, each 1 m long
photocopy (or overhead) of food-chain organisms, Teacher Table 2-1
For the teacher
overhead transparencies of Teacher Figures 2-1 and 2-2
overhead projector

Student Activity 2.4: Food-Chain Rummy

Per group of 2 to 4 students
3 sets of 15 organism cards (in resealable plastic bag), made from images on pages 25 and 26
3 sun cards, from image on page 26 (in same bag)

Student Assessment 2.5: They All Hang Together

Per student
journal entries, activity sheets, and notes from Lesson 2
Per group
animal and plant identification guide or materials
Per group, materials will vary
poster board
cardboard
markers, paint, glue, or other materials
pictures of aquatic organisms
music
costumes
computer
telecommunications system
presentation software
slides and slide projector
other presentation media

Vocabulary

adaptation
algae
bacteria
bacterium
balanced ecosystem
carnivore
closed ecosystem
consumer
decomposer
ecology
ecosystem
energy
estuarine wetlands
food chain
food web
fungi
fungus
herbivore
lacustrine wetlands
marine wetlands
nutrient
omnivore
open ecosystem
palustrine wetlands
photosynthesis
phytoplankton
plankton
primary consumer
producer
protist
respiration
riverine wetlands
scavenger
secondary consumer
tertiary consumer
trait
wetlands
zooplankton

Background for the Teacher

The material in Student Information 2.1: River and Stream Ecology summarizes, in general, the concepts that the students will encounter during this lesson. You may wish to supplement this sheet with a short lecture or additional readings on information specific to your broader biology curriculum or your local ecosystem. Many of the natural systems students will see in the field later in *Rivers Biology* are modeled in the classroom activities in Lesson 2, using an aquatic environment in an aquarium, a simulation, and a card game.

Introducing the Lesson

1. Explain that students will be taking part in activities that will introduce them to the interaction of aquatic organisms. Have the students read and discuss Student Information 2.1: River and Stream Ecology.
2. Have students complete and discuss the questions in Student Information 2.1. (Answers for student sheets are in Appendix B.)

Developing the Lesson

1. As individuals or in groups, have students carry out the procedures and observations for Student Activity 2.2: How Does Your Aquarium Rate? If possible, allow students to observe not only during class time but return later to complete the observations.
2. Have students complete Student Activity 2.2. Discuss their results and answers.

3. Arrange the classroom furniture to create an open space. Distribute Student Activity 2.3: How Does a Food Web Work?.
4. Select student roles for this activity: organisms (15), "web connectors" (several), and a student to read list. Assign the students not selected for a specific role to diagram the food chain as it is assembled.
5. Have each "organism" student pin on an organism card. Each web connector should have a piece of string. (If the class is small, eliminate web connectors, and just have the "organism" students connect the strings.) Give the reader the list of food chains shown in Teacher Table 2-1 on page 24. If desired, have the reader use an overhead of this list, revealing each new chain as it is read.
6. Have students complete procedure and observations for Student Activity 2.3. As students proceed, discuss how the string represents the flow of energy from one organism to another.
7. Have students compare their results with the overheads of the sample food web in Teacher Figure 2-1 and the lists of possible food chains in Teacher Figure 2-2. Note any chains they may have missed or added. (Students may identify legitimate relationships not included on the figure.)
8. Have students complete Student Activity 2.3. Discuss their results and answers.

Concluding the Lesson

1. Distribute and discuss Student Activity 2.4: Food-Chain Rummy, with the associated cards. Have each group play several rounds of the game.
2. Before putting the cards away, have each group sort its deck into the following categories: producer, primary consumer, secondary consumer, or tertiary consumer; herbivore, carnivore, or omnivore. Have them discuss how to label the sun in this chain.
3. Have students complete Student Activity 2.4. Review their answers.
4. Encourage the class to discuss any changes in their sense of the biological complexity of an environment based on their activities in this lesson.

Assessing the Lesson

1. To encourage creativity while developing and assessing student understanding, have the students do Student Assessment 2.5: They All Hang Together, in groups or as individuals. Clarify how much preparation of the presentations students should do as homework. If possible, arrange for videotaping of the presentations. Here is a suggested scoring rubric.

Score	Expectations
0	No presentation made.
1	Presentation is minimal. Food webs and chains did not represent actual or realistic ones. Used vocabulary words minimally, only in the simplest context, or without relevance. Most group members took part in the presentation.

2	Presentation is made. All group members took part. The three food webs and chains were complete and represented actual or realistic ones. Used vocabulary words adequately and in simple context.
3	The presentation was well done. All group members took part. Students made meaningful connections between the vocabulary and the three food chains or webs. The presentation medium was appropriate to the task and used a creative approach.
4	The presentation was well done. All group members took part. Students made excellent connections between the vocabulary and the three food chains or webs. The presentation medium was excellent for the task and used a creative approach.

2. As a simple traditional assessment, you can prepare a test for the vocabulary presented in this lesson instead of Student Assessment 2.5.
3. Have the students add to their river collages.

Extending the Lesson

1. Have the students build an aquarium of a local ecosystem and stock it with the proper organisms. For a simple approach, students may create a pond in a bottle by using cleaned two-liter bottles or glass gallon food-service jars from the cafeteria.
2. At a saltwater aquarium, have students observe the organisms, noting the relationships. Have them consider ways in which the organisms act differently from those in the aquarium at school.
3. Have students add some aquatic organisms from a local pond to the school aquarium and observe what happens over time.

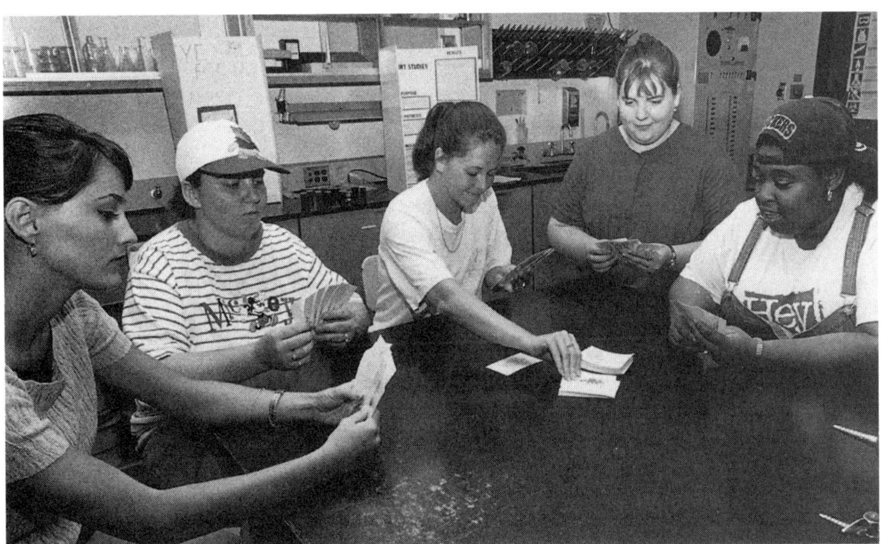

Transparency Master

TEACHER TABLE 2-1
List of Possible Food-Chain Organisms
Eagle eats large-mouth bass.
Beetle larva eats algae.
Johnny darter eats snail.
Great blue heron eats white crappie.
Red-eared sunfish eats snail.
Small-mouth bass eats flathead minnow.
Rock bass eats beetle larva.
Flathead minnow eats johnny darter.
Raccoon eats large-mouth bass.
Small-mouth bass eats johnny darter.
Great blue heron eats flathead minnow.
Snail eats eel grass.
Eagle eats red-eared sunfish.
Large-mouth bass eats rock bass.
Red-eared sunfish eats beetle larva.
White crappie eats red-eared sunfish.
Small-mouth bass eats johnny darter.
Flathead minnow eats daphnia.
Large-mouth bass eats red-eared sunfish.
Raccoon eats small-mouth bass.
Daphnia eats algae.

Eel Grass

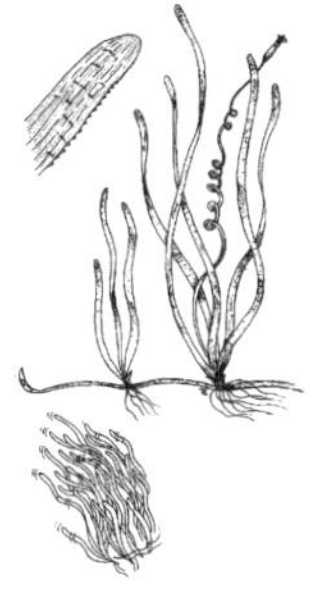

Image source: TVA, Water Management, Clean Water Initiative

Algae (Clean Water)

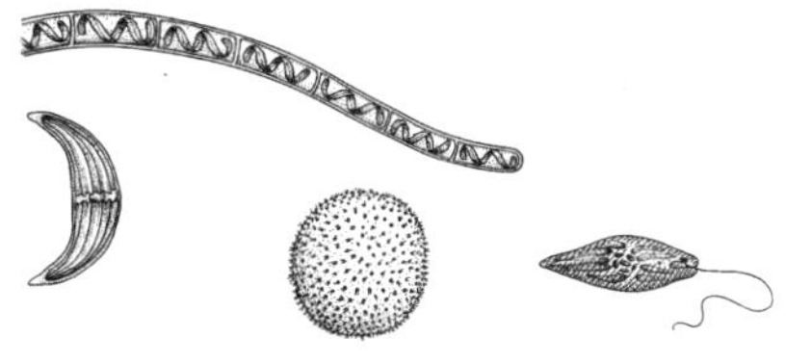

Image source: TVA, Water Management, Clean Water Initiative

Daphnia

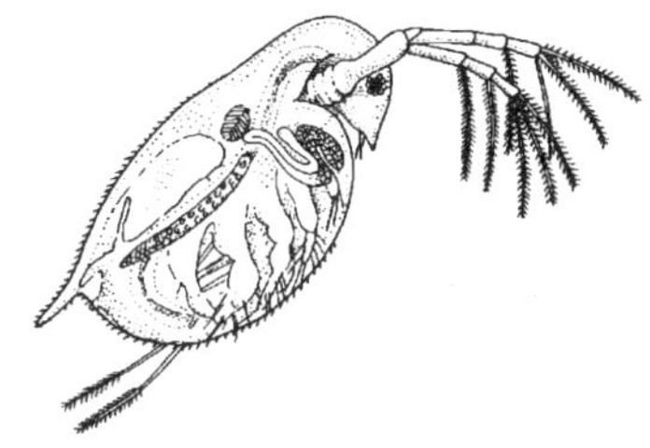

Image source: TVA, Water Management, Clean Water Initiative

Right-Hand Snail

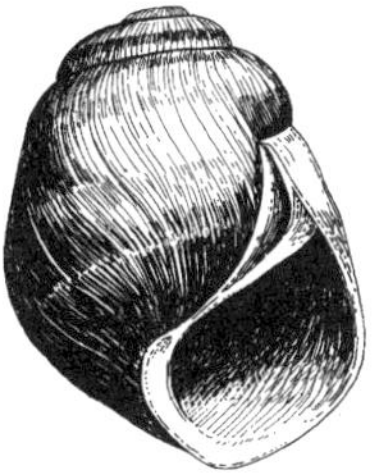

Image source: TVA, Water Management, Clean Water Initiative

Fathead Minnow

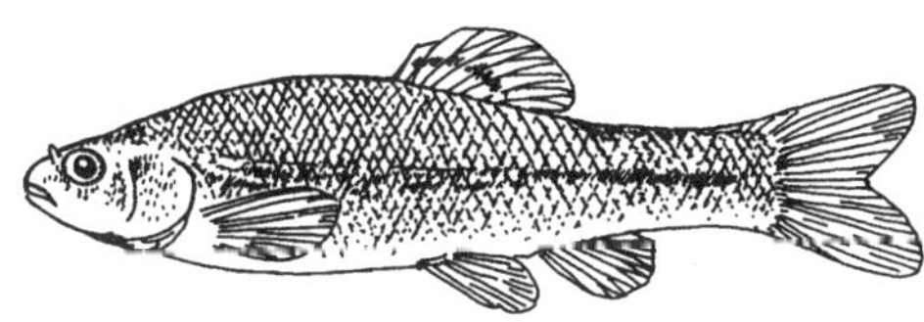

Image source: Illinois Department of Natural Resources

Johnny Darter

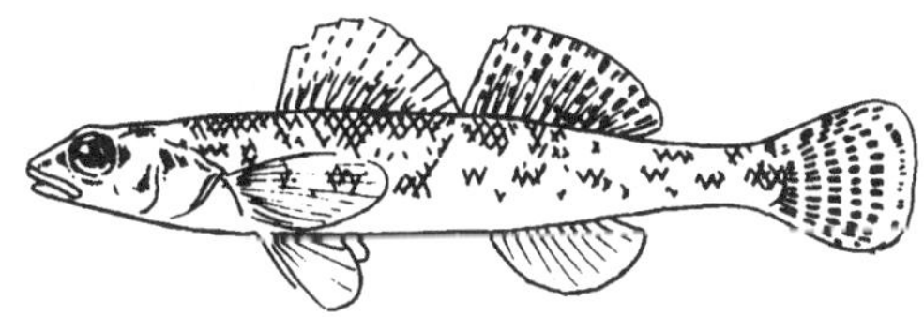

Image source: Illinois Department of Natural Resources

Great Blue Heron

Image source: Michigan United Conservation Clubs

Small-Mouth Bass

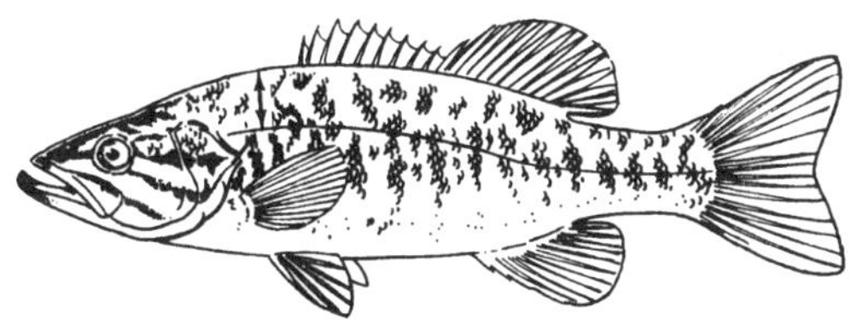

Image source: Illinois Department of Natural Resources

White Crappie

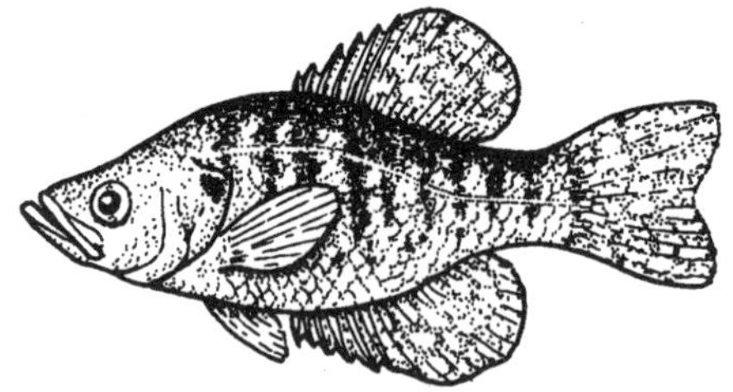

Image source: TVA, Water Management, Clean Water Initiative

Beetle Larva (Riffle Beetle)

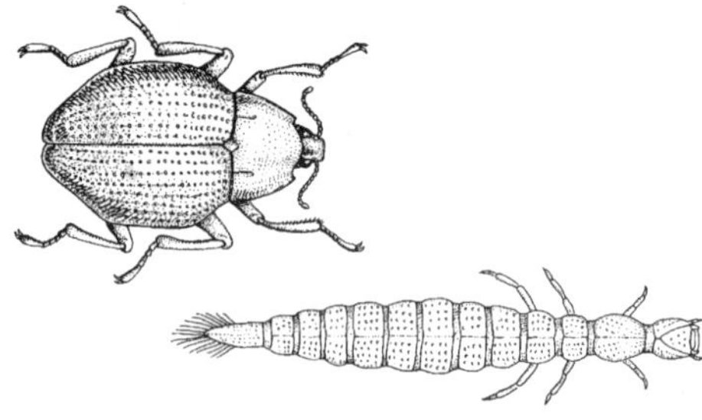

Image source: TVA, Water Management, Clean Water Initiative

Red-Eared Sunfish

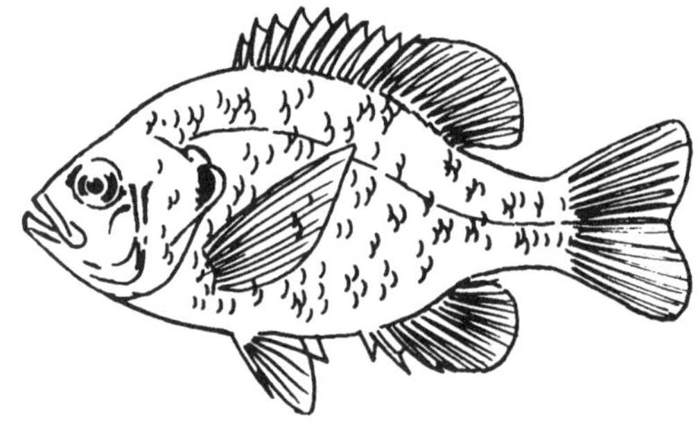

Image source: Illinois Department of Natural Resources

Bald Eagle

Image source: Michigan United Conservation Clubs

Rock Bass

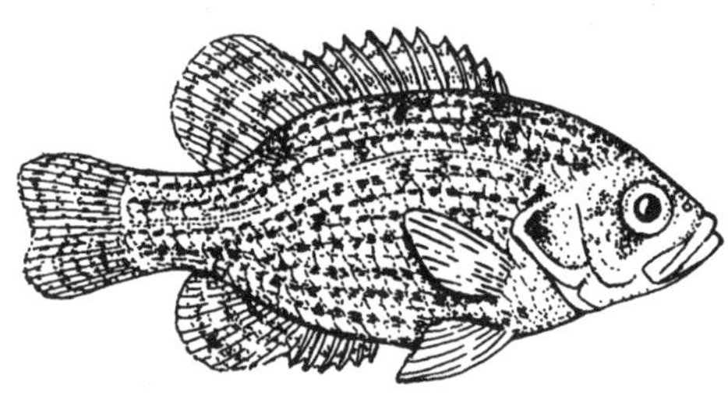

Image source: TVA, Water Management, Clean Water Initiative

Large-Mouth Bass

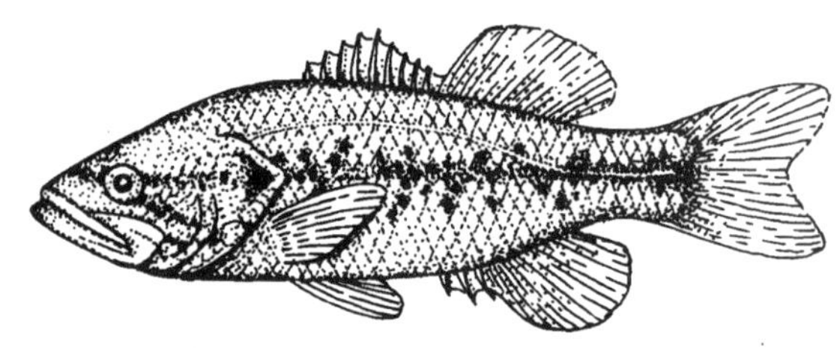

Image source: TVA, Water Management, Clean Water Initiative

Raccoon

Sun

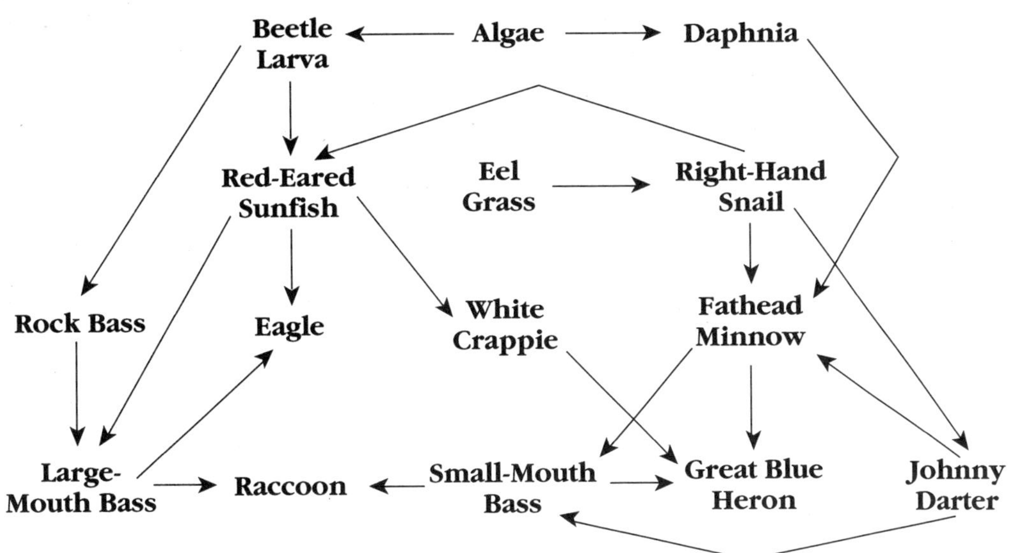

EAGLE

Eel Grass → Right-Hand Snail → Red-Eared Sunfish → Eagle

Algae → Beetle Larva (Riffle Beetle) → Red-Eared Sunfish → Large-Mouth Bass → Eagle

Algae → Beetle Larva (Riffle Beetle) → Rock Bass → Large-Mouth Bass → Eagle

RACCOON

Eel Grass → Right-Hand Snail → Red-Eared Sunfish → Large-Mouth Bass → Raccoon

Eel Grass → Right-Hand Snail → Johnny Darter → Small-Mouth Bass → Raccoon

Algae → Daphnia → Fathead Minnow → Small-Mouth Bass → Raccoon

GREAT BLUE HERON

Eel Grass → Right-Hand Snail → Johnny Darter → Fathead Minnow → Great Blue Heron

Eel Grass → Right-Hand Snail → Red-Eared Sunfish → White Crappie → Great Blue Heron

Eel Grass → Right-Hand Snail → Johnny Darter → Small-Mouth Bass → Great Blue Heron

Algae → Daphnia → Fathead Minnow → Great Blue Heron

LARGE-MOUTH BASS

Algae → Beetle Larva (Riffle Beetle) → Red-Eared Sunfish → Large-Mouth Bass

Algae → Beetle Larva (Riffle Beetle) → Rock Bass → Large-Mouth Bass

River and Stream Ecology

No living thing exists entirely by itself. The health, or quality, of a river or stream depends on interactions between and among the entities living there. A system of interactions between living organisms and their environment is called an **ecosystem.** The study of the interactions among plants, animals, and their nonliving environment within an ecosystem is called **ecology.**

Following the Energy Flow

Each organism, from the smallest plant to the largest blue whale, needs a constant source of energy, either as food or from the sun, in order to be able to grow, move, and reproduce. To obtain that **energy,** that fuel or capacity for doing activity, they all either eat regularly or engage in photosynthesis. **Photosynthesis** is the process of using sunlight to convert carbon dioxide and water into food energy, characteristic of all chlorophyll-possessing organisms. The ultimate source of the energy for most earthly organisms is the sun. (A few organisms found deep within the ground or ocean have evolved other chemical sources for energy.)

Plants and plantlike forms capture the initial energy through photosynthetic activity. Plants and plantlike one-celled organisms, called **protists,** are later eaten by bigger organisms, which are in turn eaten by larger ones. When plants and animals die and fall to the ground or the bottom of the water, they are eaten or broken down by **decomposers.** These include organisms such as **bacteria** (bak TEER ee ah)—one-celled microorganisms with no chlorophyll [singular is **bacterium** (bak TEER ee uhm)]. They also include **fungi** (FUHN guy)—mushroom-like organisms [singular is **fungus** (FUHN guhs)]. Such organisms reside in soil and water, breaking down the leaves, limbs, and trunks of wetland plants. A dead animal in the water rots and is broken down by these organisms. Decomposers release **nutrients** (food for other organisms to grow) into the soil or water as a by-product of their respiration. **Respiration** is the exchange of gases in order to obtain the compounds required for energy, specifically, using oxygen, which organisms use to break down food materials into usable materials plus carbon dioxide and water. Plants take up and use these nutrients as they capture energy from the sun and continue to grow.

Where Does a Food Chain Go?

The sequence of events in which energy is passed from organism to organism in the form of nutrients is known as a **food chain.** The smallest link of a food chain is microscopic floating or weakly swimming organisms called

plankton. Some plankton are photosynthetic plants (**phytoplankton**); others are animals (**zooplankton**).

In a stream or river, small one-celled plants (**algae**) and plantlike protists intermingle with larger plants such as duckweed, mosses, cattails, and reeds growing in the water and along the shore. In the food chain, plants are **producers,** which means that they produce their own food some of which will be passed along the chain when eaten or released when decomposed.

The second link of the food chain is animals that feed directly on plants (**consumers**). A tadpole, for example, eats algae off rocks, and a duck feeds on aquatic plants such as duckweed and algae. Such animals are called the **primary consumers** because they are the first organisms to use the energy trapped by the green plants. Some primary consumers are very small, even microscopic animal forms of plankton, known as zooplankton. (Some zooplankton can also be secondary consumers.) Primary consumers are also called **herbivores** (EHR bih vohrs), which simply means "plant eaters."

The third link in the food chain is that of the **secondary consumer,** an animal that eats primary consumers. More simply put, secondary consumers eat animals that eat plants. Because they eat animals, they are called **carnivores.** In the stream ecosystem, an example of a secondary consumer would be a turtle, since it consumes tadpoles, which are primary consumers. In this example, the chain can be extended to another organism that eats the turtle, such as a raccoon. The raccoon, like others that eat secondary consumers, would be a **tertiary** (TUHR shee ayhr ee) **consumer.**

What is a Food Web?

Any ecosystem contains many different food chains; in most cases, each food chain has at least one food source in common with others. For example, in a stream, tadpoles could be eaten by crayfish, birds, raccoons, fish, and turtles. Fish and turtles may be consumed by birds, raccoons, and humans. This overlapping and intersecting of food chains is referred to as a **food web.**

The interconnection of food chains, such as in a river or stream ecosystem, allows each animal in any food web to obtain nutrients (energy) from several food sources. Some organisms occupy different types of positions in various food chains in the food web, eating plants in one food chain, animals in another. Animals that eat both plants and animals are called **omnivores.** The term **scavenger** is sometimes given to omnivores that will eat animals that are already dead; they may also eat garbage.

Adapting to a Watery Environment

As you realize, not all species of plants and animals can live in all environments. An organism survives, and survives best, in an environment suited to its needs.

DRIFTWOOD

"We are part of this Earth and it is a part of us. All things are connected. Whatever befalls the Earth befalls the son of Earth. If men spit upon the ground, they spit upon themselves. Contaminate your bed, and one night you will suffocate in your own waste. Even the white man cannot be exempt from the common destiny. We may be brothers after all. We shall see."

Chief Seattle

One important determinant in the presence of a particular species in a specific habitat is how well it is adapted to that environment. **Adaptation** has two meanings. First, adaptation is the process of change in the physical or behavioral traits of a species due to some environmental pressure. (A **trait** is any characteristic that can be passed from parent to offspring.) This change in the species, over time, supposedly improves its ability to survive in a particular habitat.

Adaptation also relates to a chance event that occurs, thus, allowing some organisms to survive at a better rate due to some current advantage they have in a physical or behavioral trait. Organisms may have a structure, coloration, behavior, or other trait that aids them in one habitat but could be detrimental in another habitat or have no apparent function at all.

For instance, a stream whose waters have high sediment content will have animals in it that are adapted to hiding in the muddy water and escaping from the constant downward bombardment of soil particles. Imagine that humans build a dam upstream that causes the water to run clear. Though these organisms are not harmed directly by the clear water, they are exposed and haven't adapted to hiding in the clear water. Birds find them easily; soon the number of these organisms is decreased. In floats a group of organisms from upstream that can hide themselves among the plants and rocks of this clear stream. They have a body structure and behavior traits that have adapted to clear water; they flourish.

Organisms living in wet environments have adapted to a particular wetland habitat. **Wetlands** are places in the land that are covered by a shallow layer of water for some time during the year and have saturated soils from that contact. Five general classifications of wetlands are marine, estuarine, riverine, lacustrine, and palustrine.

Marine wetlands are associated with salt water and oceans. Tidal pools, mud flats, and beach areas are typical of this type. Algae are the major plants found here. **Estuarine** (ES chew reen) **wetlands** are those that occur where fresh water and salt water mix. This occurs at the mouths of rivers that flow into the ocean. Mangrove swamps form one type, as do the sawgrass tidal areas of the Southeastern United States.

Riverine wetlands are found with moving water along rivers and streams. **Lacustrine** (LA kuh strinn) **wetlands** are associated with ponds, lakes, and reservoirs, which have slow water movement. These two wetlands types have the greatest number of organisms associated with them. The final classification of wetlands is **palustrine** (PAL uh strinn). This type is made up of the bogs, marshes, and wet meadows found in forests and prairies. The Arctic tundra is filled with many such wetlands areas. The prairie pot hole is an example of one of these wetlands.

Several types of wetlands may exist in the same area or lay adjacent to each other. A riverine environment in the north woods can have ponds and lakes

DRIFTWOOD

"As scientists understand more and more about the interdependence not only of living things but of rocks, rivers—the whole of the universe—I am left in awe that I, too, am a part of this tremendous miracle. Not only am I a part of this pulsating network, but I am an indispensable part. It is not only theology that teaches me this, but it is the truth that environmentalists shout from the rooftops. Every living creature is an essential part of the whole."

Desmond Tutu
Anglican Archbishop of South Africa

nearby, as well as a bog or marsh. Some species inhabit only one of those habitats; others can survive in numerous settings. The great blue heron and cattail are adapted to all of those three wetland types, but a pitcher plant or a sundew survives only in a marsh. Generally, the more mobile the organism, the more types of wetlands in which it is adapted to live. Wetlands can be generally identified by the major plants found there.

As you observe plants and animals in your local stream or river environment, pay attention to who eats what and how they grow. Understanding the food chains and food web in that habitat is key to protecting and nurturing each species. Pay attention, also, to how the characteristics of the organisms you observe are adaptations to such a setting.

Questions

1. Explain the difference between food chains and food webs.
2. Where does energy come from? How is it captured so organisms can use it?
3. Explain the difference between primary consumers and secondary consumers. Provide a few specific examples. What are humans?

   Use Figure 2-1 to answer Questions 4 and 5. In the figure, each number represents an animal in a food chain or web.

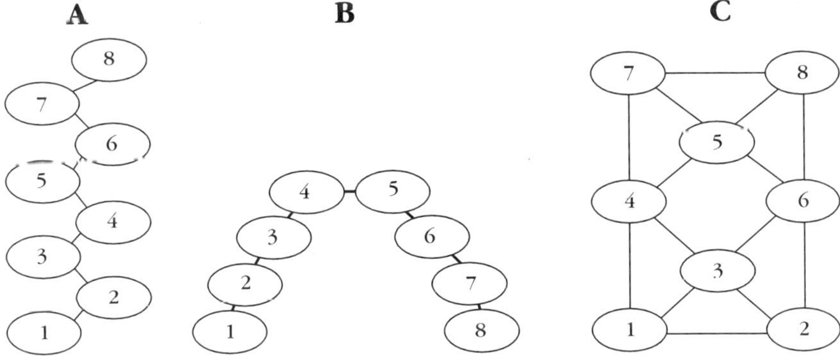

Figure 2-1: Sample Food Chains and Web

4. What would happen to each structure if object number 3 were removed? Why?
5. If each number represents an organism in a food web within an ecosystem, which system would most likely endure over time? Why?
6. Look at all the new words printed in bold in this information sheet. Review their definitions, as they are used in the study of wetlands. Select at least five words that are new to you and write a paragraph that includes those words.

Name

STUDENT ACTIVITY 2.2

How Does Your Aquarium Rate?

Purpose

To observe and describe interactions taking place in an aquatic ecosystem.

Background

You will observe an aquatic ecosystem to determine the types of food chains and food webs it contains. NEVER tap or pound on the glass of the aquarium to get the attention of the animals. Some fish are so disturbed by loud noise that they may go into shock.

The aquarium that you are observing may be an **open ecosystem** or a **closed ecosystem.** An open ecosystem is open to the outside world, allowing materials to enter or leave. In an open aquarium, for instance, food is added, gases leave or are replaced by an aerator, and new animals or plants are exchanged. The **closed ecosystem** is sealed off from the outside. Only enough plants are present to provide food and oxygen for the animals. The animals feed and fertilize the plants. They also provide carbon dioxide for the plant. If an animal or plant dies, it cannot be replaced in a closed system. The closed system that is functioning properly is said to be a **balanced ecosystem.**

You will observe a balanced aquarium.

Materials

Per two students

- medicine dropper
- depression slides
- microscope

Per group or class

- balanced aquarium, open (assembled 4 weeks earlier), containing:
  live fish
  live plants
  live snails
  aquarium gravel
  other appropriate organisms
- identification books showing microscopic aquatic organisms

Procedure

1. Sit some distance from the aquarium. Observe the organisms. In your journal (or on a blank piece of paper, if instructed by your teacher), draw the aquarium in both side view and top view. In your drawing, include all the aquatic plants (producers) and living animals (consumers). If any organisms are moving, draw them in the area where they are found most of the time. Beside the drawing, make a list of all the organisms in the aquarium.
2. List all the organisms you observe, giving as specific a name as possible.
3. Move closer to the aquarium. Observe the moving organisms in the aquarium for at least 5 minutes. Record interactions between organisms. Use arrows to show consumer relationships.
4. Based on your observations, draw the food web of your aquarium.
5. Using a medicine dropper, obtain an aquarium water

sample from near the top of the tank. Disturb the ecosystem as little as possible. Use a microscope slide and cover slip to make a wet mount of the sample. Observe the sample under a microscope. In your journal, note and draw your observations. If you want to look at a variety of animals, examine under the microscope the material in the filter, because it probably houses many types of organisms.

6. Repeat Step 5 for two other water levels, such as midway and near the bottom of the tank.
7. Make drawings in your journal of the organisms found in each of the three water samples. If possible, identify and label the microorganisms.

Observations

1. Include your side-view and top-view drawings of the aquarium.
2. Include your diagram of the aquarium food web.
3. Include your drawing of each microscope sample, identifying specific organisms when possible. Use Figure 2-2 as a model.

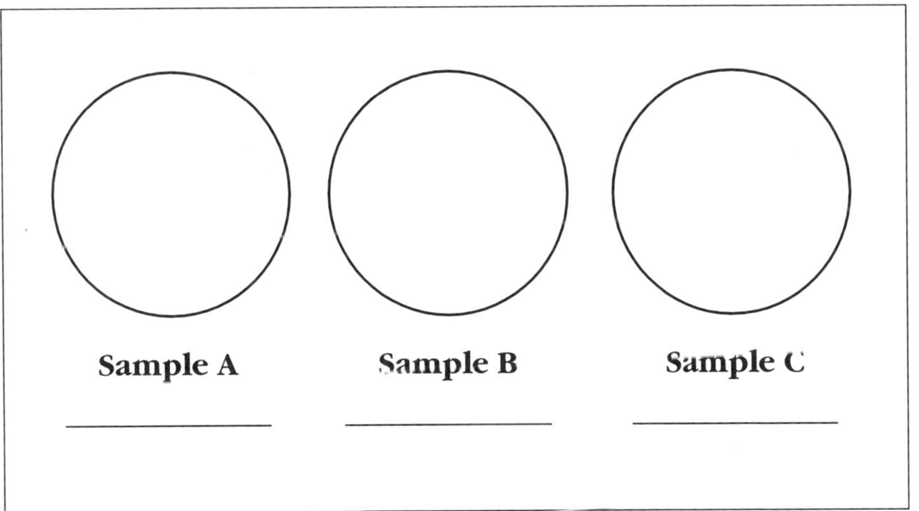

Figure 2-2: Model for Microscope Drawing

Analysis and Conclusions

1. What is the most plentiful organism visible when you observed the tank? Why do you think this is?
2. What would happen if you increased the number of fish in your aquarium?
3. How do the organisms found in each drop of water fit into the aquarium food chain?
4. Knowing that fish eat invertebrates, explain why you saw or did not see invertebrates in your aquarium water samples.
5. If members of your class observed more than one aquarium, compare observations with those from other groups. Build a table that compares the observations. Make hypotheses for any differences observed among the tanks.

Critical Thinking Questions

1. Would you recommend any additional organisms or types of organisms for this aquarium? Explain your reasoning.
2. How is your aquarium different from a natural pond?

Keeping Your Journal

1. What can you do to the aquarium you observed to make it more like a natural aquatic environment?
2. If you were an aquatic organism, would you like to live in this aquarium? Why or why not?
3. If you have an aquarium at home or have access to an aquarium, observe the organisms in that system. Note the relationships. Discuss whether these organisms act differently from those at school and, if so, why.

Name

STUDENT

ACTIVITY 2.3

How Does a Food Web Work?

Purpose

To understand how food chains form a food web.

Background

Your class will simulate how food chains become interconnected to form a food web. This activity models a common food web that exists in many aquatic environments. You will observe more complex webs as your knowledge builds.

Procedure

1. As your teacher directs, move furniture to make a large open area.
2. Your teacher will assign roles. If your teacher assigns you to be an organism, select one organism card and pin it on to identify yourself.
3. If you will be a web connector, take several strings provided by your teacher.
4. If you will be the reader, your teacher will give you a list of food chains.
5. If you have no special assigned identity, you will be responsible for drawing the food web as it is formed in your classroom. Label each organism. Use arrows to indicate connections, with the energy flow to the consuming organism. For example, if the reader says, "Great blue heron eats fathead minnow," the arrow should point from the fathead minnow toward the heron, because energy moves from the fathead minnow into the eagle as it eats, as shown in Figure 2-3.
6. From the list of food chains, the reader will give the name of an organism and then the name of its consumer. If you represent either of these organisms, step into the open space. If you are a web connector, use string to connect the organism to its consumer. Tie a wrist of one student to a wrist of the other, or have them hold the string. Those drawing the

Fathead Minnow ⟶ Great Blue Heron

Figure 2-3: Sample Food-Web Connection

Materials

Per student

- colored pencils or markers

Per class

- set of 15 organism cards (in resealable plastic bag), made from images on pages 25 and 26
- 15 safety pins (or pieces of masking tape)
- 20–25 pieces of string, each 1 m long
- photocopy (or overhead) of food-chain organisms, Teacher Table 2-1

For the teacher

- overhead transparencies of Teacher Figures 2-1 and 2-2
- overhead projector

food web should diagram this relationship.

7. Repeat Step 4 until reader has read all food chains, and all web connections are made. Hold the web together long enough that assigned students can finish drawing the web model.
8. Dismantle the food web. Return materials to your teacher.

Observations

9. Students who have drawn the food web should share their sketches with the class. Individually or in groups (as instructed by your teacher), draw, in your journals, the model food web. Label each organism. Use arrows to indicate interconnections in the web, with the energy flow to the consuming organism.

Analysis and Conclusions

1. How many different food chains were part of this food web? List each chain, or use colored pencils to trace each chain on your diagram.
2. Which organism has the most connections to it? Offer a hypothesis that would explain this phenomenon.
3. Are some organisms found in every chain? What would happen to the food web if these organisms were not present?

Critical Thinking Questions

1. Of all the organisms in this food web, whose death would impact the web the most? Support your answer.
2. Something is wrong with this food web; it does not contain any decomposers. Put them into your web diagram where you think they belong.

Keeping Your Journal

1. Using five new river animals, add another food chain to the web you have drawn. Use a different colored pencil to draw this.
2. Write a story about an imaginary watery planet. Create the organisms found there and develop a food web.
3. If you can, visit a pond, river bank, or lake, and see whether you observe any of the organisms in this activity. If you do, note their behavior in your journal.

Food-Chain Rummy

Purpose

To become familiar with some of the simple food chains that exist in most wetlands and riverine ecosystems.

Background

In this activity, you will play a food-chain card game that will reinforce what you learned in Student Activity 2.3: How Does a Food Web Work?, in which you explored the relationship between food chains and a food web. In this card game, you'll work with food chains that appear to be linear. As you play, you will see how some of these food chains might fit together to form a food web.

Materials

Per group of 2 to 4 students

- 3 sets of 15 organism cards (in resealable plastic bag), made from images on pages 25 and 26
- 3 sun cards, made from image on page 26 (in same bag)

Procedure

1. The object of the game is to assemble a hand of six cards that form a single food chain. The cards in this deck can combine to make any of the five food chains shown in Figure 2-4.
2. Shuffle the cards well and deal five cards to each player. Place the leftover cards facedown on the table. Turn the top card faceup next to the pile. These are the draw pile and the discard pile.
3. Play starts on the dealer's left and goes clockwise. The first player picks up the top card in either the draw or the discard pile. The player examines the card to decide whether it will go with the rest of the cards in his or her hand toward forming one of the food chains in Figure 2-4. Then the player must discard one card (except at the point of winning).
4. Play proceeds in turn.
5. If the draw pile runs out during the game, keep the last discard faceup. Shuffle the remaining discard cards and place them facedown as the new draw pile. Continue play.
6. To win, all six cards in your hand (counting the most recent card you picked up) must belong to one of the five food chains. To win, you must lay down the cards in the order shown on the food-chain list.
7. Write down the winning food chain in your observation data table.
8. Repeat the game several times as directed by your teacher.

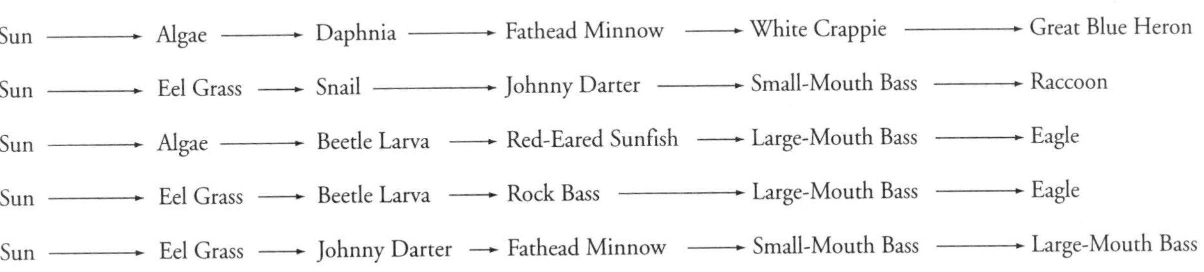

Figure 2-4: Five Food Chains in Food-Chain Rummy

Observations

DATA TABLE: Winning Food Chains

Your Group

Other Groups

Analyses and Conclusions

1. Share results with other groups. Did some food chains appear more often in winning hands? Offer an explanation for your observation.
2. After playing a few times, list at least one of the food chains from memory, with each organism in order.

Critical Thinking Questions

1. What did you learn from playing Food-Chain Rummy?
2. Is any part of the food chain more important than others? Why or why not?
3. List which organisms on the cards are producers, primary consumers, secondary consumers, and tertiary consumers. Then list herbivores, carnivores, and omnivores.
4. Change some part of one of the chains. Offer support for your choice. (For example, a brassy minnow could eat a small-mouth bass. The supporting argument might be that the bass was very young and small.)

Keeping Your Journal

1. Using one of the food chains in Figure 2-4, write a suspense story that shows what really happened in the water as that food chain took place.
2. Add yourself, a human, to one of the food chains in Figure 2-4. Tell where you would fit and why.

Name

STUDENT ASSESSMENT 2.5

They All Hang Together

Purpose

To develop a presentation that demonstrates an understanding of food webs and food chains.

Introduction

You have been introduced to the ideas of food webs and food chains in aquatic environments and have studied many basic terms common to any study of the environment. Now you will demonstrate this knowledge by using those words and concepts in a presentation about food webs and food chains in a specific type of wetlands environment.

Procedure

1. Either as an individual or in a group, from among the types of wetlands listed in Student Information 2.1: River and Stream Ecology, select one to investigate and present: marine, estuarine, riverine, lacustrine, or palustrine.
2. Use identification guides, the Internet, and any other available material to list the animals and plants that live in your wetlands.
3. Using your knowledge of those organisms, diagram at least three food webs or food chains likely to occur in that ecosystem.
4. Assemble a vocabulary list from the activities in Lesson 2. Include a definition for each word.
5. Decide on a way to present your food chains or webs to the class. Depending on available resources and on your teacher's directions, options may include: a skit, rap or other popular song format, a formal presentation, a computer presentation, a musical, an opera, or even mime.
6. Organize the structure of your presentation, including at least three food webs or chains. Create the dialog or other spoken or visual use of words, applying correctly terms introduced in Lesson 2 of *Rivers Biology*. Gather the materials you will use. Prepare any visual support for your presentation. Practice your presentation. Plan to have every member of the group take part in the presentation. If possible, arrange for someone to take pictures or preserve a portion of the presentation in some way so you can use it in your portfolio.
7. Make your presentation to the teacher and the class.

Materials

Per student

- journal entries, activity sheets, and notes from Lesson 2

Per group

- animal and plant identification guide or materials

Per group, materials will vary

- poster board
- cardboard
- markers, paint, glue, or other materials
- pictures of aquatic organisms
- music
- costumes
- computer
- telecommunications system
- presentation software
- slides and slide projector
- other presentation media

Performance Criteria

- Present organisms and food webs or food chains reasonable for the wetlands environment portrayed.
- Incorporate appropriate terms introduced in *Rivers Biology* relevant to food webs and chains.
- Demonstrate appropriate understanding of food chains and webs, and of the wetlands environment being presented.
- Creative presentation, such as use of drama, visual media, or other appropriate approach, to communicate this information to an audience effectively.
- Work cooperatively in groups and share information, planning, and responsibility in the preparation and in the presentation.

Indices of River and Stream Water Quality

Focus

Students will study the concept of an index as a way of comparing conditions of aquatic habitats. They will learn how to develop an index and become familiar with several indices useful in evaluating water quality.

Learner Outcomes

Students will:

1. Understand how indices are used.
2. Understand the significance of water-pollution tolerance and diversity indices.
3. Recognize common types of benthic macroinvertebrates used for evaluating water quality.
4. Be able to determine water quality using index values.

Time

Three to four class periods of 40–50 minutes per period

DAY 1: Student Activity 3.1: Creating and Using an Index

DAY 2: Student Information 3.2: Biological Indices of Water Quality
Student Information 3.3: Benthic Macroinvertebrates and Pollution Tolerance

DAY 3: Student Activity 3.4: The Water-Quality Index

Advance Preparation

Prepare to supply students with Student Information and Activity sheets 3.1 through 3.4. Gather all necessary materials for this lesson. Make arrangements for each group to have a calculator with a square-root function. Prepare overhead transparencies of Teacher Table 3-1: Model of Water-Quality Index. For each group, prepare and laminate one set of benthic macroinvertebrate cards, using photocopies of the images on pages 45 through 50. Set them up as two-sided cards, with a picture on one side, the corresponding information on the reverse.

Become familiar with the method used for determining each index presented in this lesson. As well as reading the materials presented here, you may want to contact state agencies to discuss the counterpart indices they use. A local conservation agency may be able to help, even come work with students to complete the indices. Also refer to organizations and other resources in Appendix C, such as The Izaak Walton League; this group has been the leader in water monitoring and has good materials on benthic macroinvertebrates. Begin a collection of identification books, resource materials, and other resources such as those listed in that appendix.

For more background on the water-quality index discussed in Student Activity 3.4, see *Rivers Chemistry*.

Materials

Student Information 3.3: Benthic Macroinvertebrates and Pollution Tolerance

Per group

set of laminated benthic macroinvertebrate cards based on the images on pages 45 through 50

Student Activity 3.4: The Water-Quality Index

Per group

calculator

For the teacher

overhead transparency of Teacher Table 3-1, Model of Water-Quality Index

overhead projector

Vocabulary

abiotic
benthic macroinvertebrate
benthic macroinvertebrate pollution tolerance index
biotic
detritus
diversity index
flow rate
index
indices
invertebrate
limiting factor
macroinvertebrate
macrophyte
microinvertebrate
organic matter
overall Water Quality Index (WQI)
periphyton
protocol
Q-value percent
riffle
substrate
taxa
taxon
tolerance
weighting factor

Background for the Teacher

The indices included in Lesson 3 provide the statistical structure around which students can learn how water researchers assess and compare water quality. Determining these indices involves use of data on biological and chemical characteristics of river and stream water. Later in *Rivers Biology,* field collection and identification of specific organisms as well as data from chemical tests of water samples will provide data students can use to determine specific water-quality indices.

The first biological index for evaluating river and stream water quality was a Pollution Tolerance Index produced and promoted by the Save Our Streams (SOS) program of the Izaak Walton League. Many states have since created such indices for volunteer monitors. *Citizen Stream Monitoring: A Manual for Illinois* (1990) and *Ohio's Stream Quality Monitoring: A Citizen Action Program* (1990) are two such guides. The index used in this activity, based on the Illinois model, is designed for nonexperts. Scientists have also created many more complex indices. These indices evaluate specific traits or are based on taxonomics that identify organisms to family or species.

Though species may vary somewhat in different areas, the general types of benthic macroinvertebrates found at specific pollution levels are similar. The indices included in this lesson are designed to provide monitoring data for use universally. If desired, however, you may obtain from your state environmental agencies an index for citizen monitoring tailored to your area or correlated to water-quality data already available. Once students have identified the macroinvertebrates in the river or stream water and noted the taxa in their journals, they can easily apply the data to more than one index.

Introducing the Lesson

1. Have students read, discuss, and answer the questions for Student Activity 3.1: Creating and Using an Index. (Answers for student sheets are in Appendix B.)
2. Discuss what an index does and give some examples of ones used in the classroom. Have students brainstorm a list of indices that our society uses regularly, such as ACT, SAT, mold counts, and air-pollution indices. Keep (and perhaps post) the list, so students can add to it later if desired.

Developing the Lesson

1. Have the students read, discuss, and answer the questions for Student Information 3.2: Biological Indices of Water Quality.
2. Have the students read, discuss, and answer the questions for Student Information 3.3: Benthic Macroinvertebrates and Pollution Tolerance.
3. Distribute the laminated benthic macroinvertebrate identification cards. Have students work in groups, using these as flash cards to practice recognition and tolerance grouping for these organisms. If you have additional pictures of benthic macroinvertebrates to supplement those in *Rivers Biology,* share them with the students.

Concluding the Lesson

1. Have the students read and discuss Student Activity 3.4: The Water-Quality Index (adapted from *Rivers Chemistry*). Work through the mathematical steps as a class. Some of the tests needed in order to produce this index are included in *Rivers Biology;* for teaching activities about the balance of the tests, see *Rivers Chemistry*. As appropriate, use an overhead transparency of Teacher Table 3-1, Model of Water-Quality Index, located at the end of the Teacher Notes for Lesson 3, to show students how to enter data, determine Q-value, and calculate overall water quality.
2. Have students answer questions on Student Activity 3.4. Discuss results.

Assessing the Lesson

1. Have students write a short paragraph summarizing each index.
2. Have students develop a short poem about one of the indices they have studied. For specific teaching activities on writing poetry, see *Rivers Language Arts*.

3. Real assessment for this lesson will come later in this unit when students collect data in the field and apply these observations to determining indices. At that time, assess whether, when given real data, students can complete the forms and compute the values for each index.
4. If students are doing a collage on river and stream water quality, have them add information and pictures that relate indices.

Extending the Lesson

1. Have students access the Rivers Project home page on the World Wide Web to see what data are available and which indices are on the home page.
2. Have students contact a local environmental agency, university, or other business to investigate what indices they use to report and compare data.
3. Flood level is an index for flooding. Have students find out about the flood-level index for your local area.
4. Have students contact the local weather bureau or television weather forecaster to find out what indices are used in meteorology, including how they were developed.

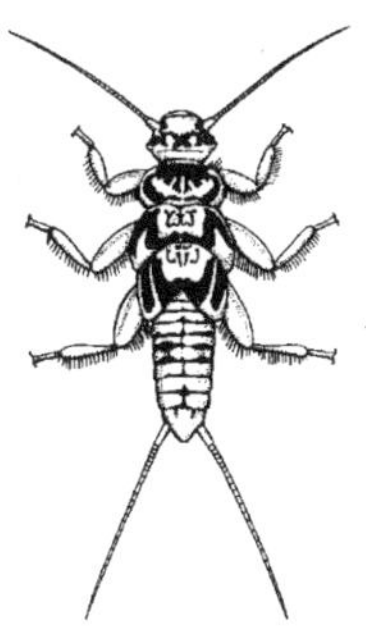

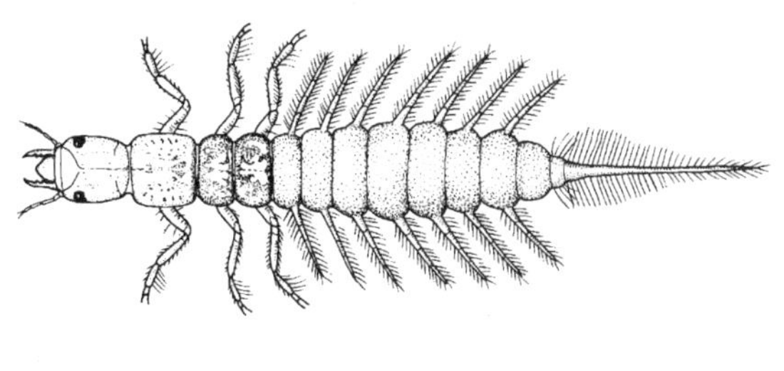

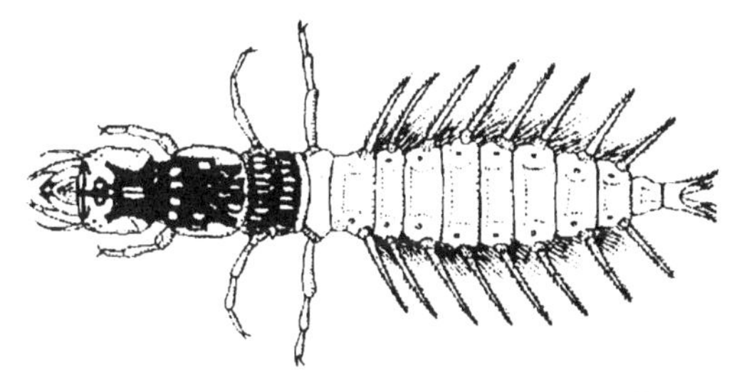

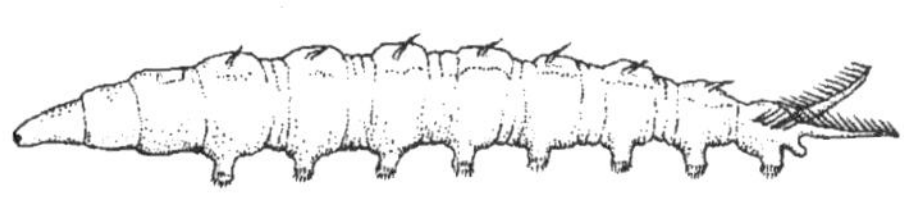

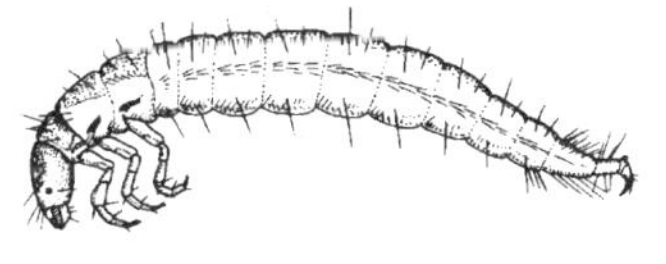

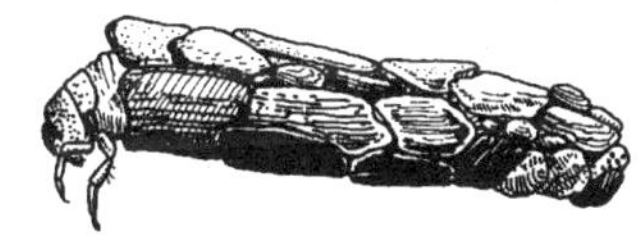

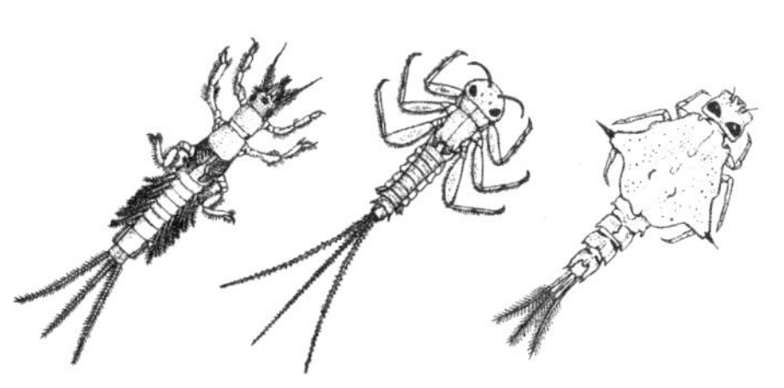

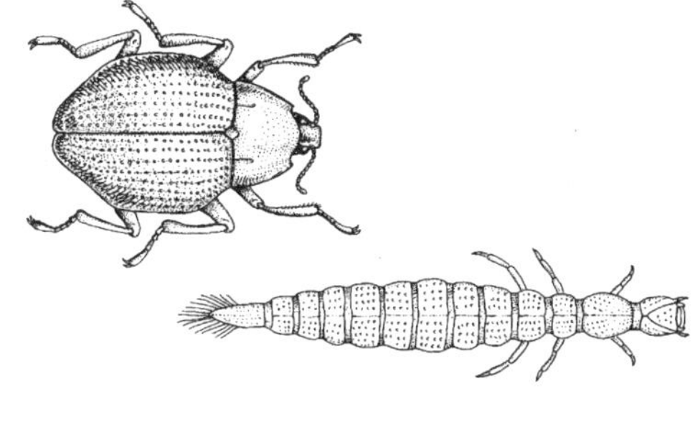

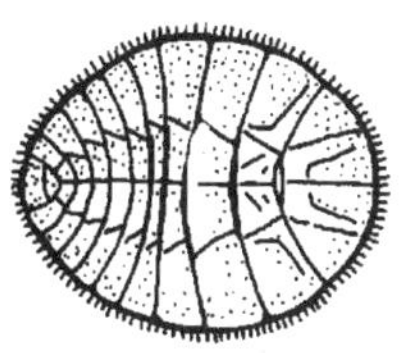

ALDERFLY
(Order Megaloptera, Family Sialidae)

Description: Adult: Length to 25 mm. Dark with large wings folded over the body. Larva: Length to 30 mm. Body is thick-skinned with 6–8 filaments on each side of the abdomen, plus a single tail filament. Gills at the base of each side filament. Color brown. *Reproduction:* Female deposits eggs on vegetation over the water. Larva hatches into the water. *Food:* Larva is predator on other aquatic invertebrates. Fed on by other larger predators. *Pollution Tolerance:* Intolerant.

Image source: TVA, Water Management, Clean Water Initiative

STONEFLY
(Order Plecoptera)

Description: Adult: Length to 25 mm. Resembles nymph but with two long pairs of wings folded over body. Two cerci. Color drab grayish and whitish. Nymph: Length up to 50 mm. Two distinctive tails, cerci. Brightly colored in tan, brown, gold, and black. Sluggish movement in water. *Reproduction:* Most females deposit eggs on top of water and they drift to the bottom. *Food:* Some are carnivorous; others feed on larvae, algae, bacteria, or plant debris. Food for a wide variety of aquatic animals. *Pollution Tolerance:* Intolerant.

Image source: TVA, Water Management, Clean Water Initiative

SNIPEFLY
(Order Diptera, Family Athericidae, Atherix variegata)

Description: Adult: Length to 18 mm. Larva: Length to 18 mm. Elongated, cylindrical, slightly flattened. Cone-shaped abdomen with two long, fringed filaments. Color varies. *Reproduction:* Female deposits eggs on overhanging vegetation and immediately dies, attached to the egg mass. Larvae hatch and drop into the water. Adults and larvae are predaceous, and voracious on aquatic invertebrates. Not important as a food source for other animals. *Pollution Tolerance:* Intolerant.

Image source: TVA, Water Management, Clean Water Initiative

DOBSONFLY
(Order Megaloptera, Family Corydalidae, Corydalus cornutus)

Description: Adult: Length to 10 cm, wingspread to 12.5 cm. Two pairs of long, colorful wings folded back over body length. Male, long mandibles. Female, smaller mandibles. Larva: Length to 90 mm. Called hellgrammites. Large mandibles, filaments located along side of abdomen. Pair of tail filaments for grasping. Color brownish to black. *Reproduction:* Female attaches eggs to overhanging vegetation. Hatched larvae fall into the water. *Food:* Predaceous larva feeds on other aquatic invertebrates. Fed on by many fish. *Pollution Tolerance:* Intolerant.

Image source: TVA, Water Management, Clean Water Initiative

MAYFLY
(Order Ephemeroptera)

Description: Adults: Length to 25 mm. Soft-bodied with lacy wings folded upright. Front legs held forward. Usually two long cerci. Mouth parts undeveloped. Long slender antennae. Nymph: Length to 25 mm. Three cerci; occasionally two, but never paddle- or fan-like. Dark green, brown, gray, or black. *Reproduction:* Eggs deposited on top of water, drift to bottom. Some species attach eggs to submerged objects. *Food:* Algae, bacteria, plant debris, plankton, plant parts. Adults never eat. Food for fish and invertebrates. *Pollution Tolerance:* Most intolerant to moderately intolerant. Families Caenidae and Siphlonuridae, fairly tolerant.

Image source: TVA, Water Management, Clean Water Initiative

CADDISFLY
(Order Trichoptera)

Description: Adult: Length to 25 mm. Long slender legs. Moth-like. Sloping wings covered with hairs. Color tan, brown, or gray. Larva: Length to 25 mm. Wormlike with legs. May build case from stones or plant parts. Color green; may be yellow to brown. *Reproduction:* Eggs in a gelatinous mass; attached to submerged objects. *Food:* Larva eats algae, bacteria, plant debris, plankton, plant parts. Some build nets to capture drifting food. Food for fish and predaceous invertebrates. Adults never eat. *Pollution Tolerance:* Most families intolerant; Polycentropodidae and Molannidae, moderately intolerant.

Image source: TVA, Water Management, Clean Water Initiative

WATER PENNY
(Order Coleoptera, Family Psephenidae, Psephenus sp.)

Description: Adult: Length to 7 mm. Oval, flattened body. Resembles riffle beetle but can be found on rocks out of the water. Long legs and hairy body. Larva: Length to 12 mm. Resembles circular encrustation on rocks. Head and legs hidden by body. Shield-like. Color tan or brown. *Reproduction:* Female crawls into fast-moving water and deposits eggs on undersides of stones. *Food:* Primarily algae, bacteria, or plant debris. *Pollution Tolerance:* Moderately intolerant.

Image source: TVA, Water Management, Clean Water Initiative

RIFFLE BEETLE
(Order Coleoptera, Family Elmidae and Dryopidae)

Description: Adult: Length to 6 mm. Crawl on rocks in fast-moving water. Color dark brown to black. Adults aquatic and immobile. Larva: Length to 12 mm. Resembles small torpedo with circular stripes or rings around the body, six legs obvious. Pointed at both ends with a fuzzy mass at one end. Color usually grayish. *Reproduction:* Female deposits eggs on plant materials under water in fast-moving water. *Food:* Primarily plant material such as diatoms and algae, also plant debris and bacteria. *Pollution Tolerance:* Moderately intolerant.

Image source: TVA, Water Management, Clean Water Initiative

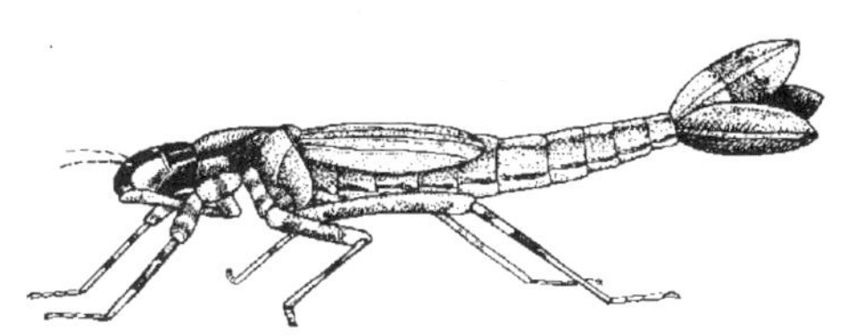

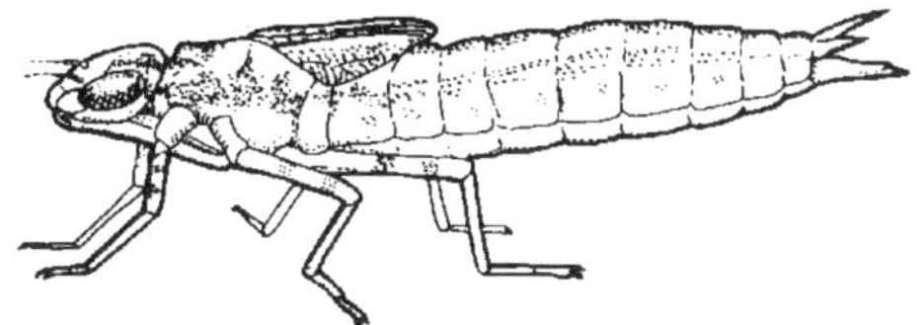

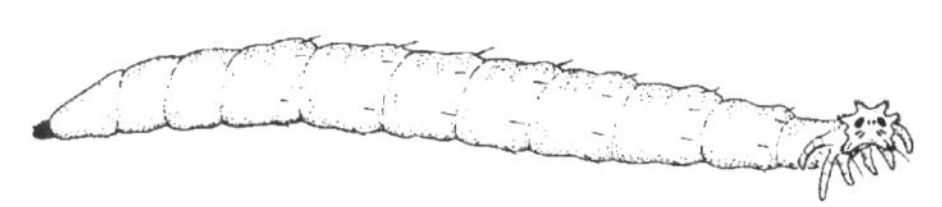

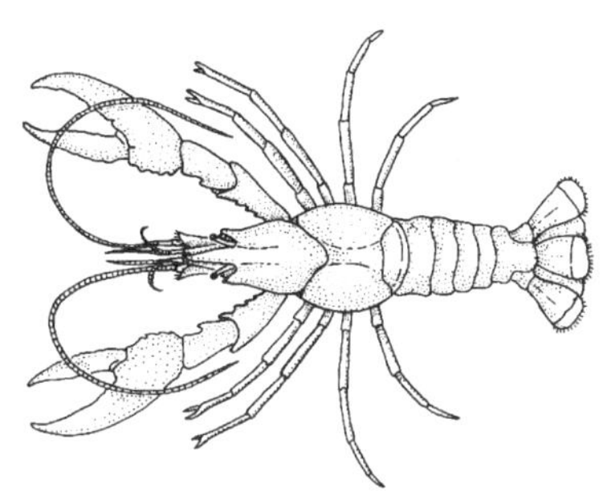

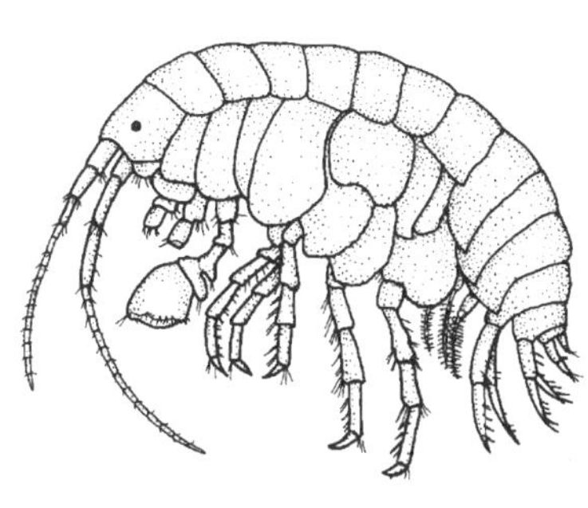

DRAGONFLY
(Order Odonata, Suborder Anisoptera)

Description: Adult: Length to 90 mm. Very long abdomen. Two pair bright wings horizontal at rest. Nymph, called naiad: Length to 60 mm. Usually robust, elongated, or spider-like body; six legs at side. Large eyes. Brown, black, gray; often green, with algae growing on surface. *Reproduction:* Flying female deposits eggs on water; eggs float down. *Food:* Feeds on invertebrates, small fish, tadpoles. Food for larger aquatic animals, mainly fish. *Pollution Tolerance:* Family Libellulidae, fairly to very tolerant. Family Gomphidae, intolerant. Families Aeschnidae, Macromiidae, Cordulegastridae, Calopterygidae, and Corduliidae, moderately intolerant.

Image source: TVA, Water Management, Clean Water Initiative

DAMSELFLY
(Order Odonata, Suborder Zygoptera)

Description: Adult: Length to 25 mm. Long abdomen; two pairs of wings rest over the back. Color bright green, blue, or red. Nymph or naiad: Length to 30 mm. Body elongated, with 3 paddle-like tails (gills) at tip of abdomen. Large eyes. Six large legs. Color varies green, brown, and black. *Reproduction:* Flying female deposits eggs on cuts in stems of aquatic plants. *Food:* Feeds on invertebrates. Food for larger aquatic animals, especially fish. *Pollution Tolerance:* Family Coenagrionidae, fairly to very tolerant. Family Lesidae, fairly to very tolerant. Family Calopterygidae, moderately intolerant.

Image source: TVA, Water Management, Clean Water Initiative

CRAYFISH
(Phylum Arthropoda, Class Crustacea, Order Decapoda, Family Cambaridae, Cambarus sp.)

Description: Length up to 15 cm. Resembles small lobster. Four pairs of walking legs and a strong set of pinchers. Color brown, green, reddish, or black. *Reproduction:* Female carries eggs and very young on swimmerets under the abdomen. *Food:* Scavenger on plant and animal debris. Pinchers hold the food for eating. Fed upon by larger predators. *Pollution Tolerance:* Moderately intolerant.

Image source: TVA, Water Management, Clean Water Initiative

CRANEFLY
(Order Diptera, Family Tipulidae)

Description: Adult: Length up to 5 cm. Looks like huge mosquito. Long slender legs. Color brownish to white. Larva: Length up to 5 cm. Worm-like, thick-skinned, brown-green to transparent whitish. Pointed at one end, with a set of disk-like spiracles at the other. *Reproduction:* Female deposits eggs on submerged vegetation or other debris. *Food:* Adult: nothing or nectar. Larva: Scavenger on organic debris. Some are predaceous. Fed on by other larger aquatic animals. *Pollution Tolerance:* Moderately intolerant.

Image source: TVA, Water Management, Clean Water Initiative

BLACKFLY
(Order Diptera, Family Simulidae)

Description: Adult: Length to 5 mm. Small, humpbacked, black, having iridescent wings. Larva: Length to 8 mm. Small, worm-like and bulbous at one end. Out of water, they fold in half while wiggling. Color white to gray *Reproduction:* Lay eggs on stones or debris in rapid water; larva and pupa also found in rapid water. *Food:* Larva eats organic debris. Adult female feeds on blood of warm-blooded animals. Larvae and flies fed upon by both aquatic and terrestrial animals. *Pollution Tolerance:* Fairly tolerant.

Image source: TVA, Water Management, Clean Water Initiative

FRESHWATER CLAM/MUSSEL
(Class Bivalvia, Family Unionidae)

Description: Great variation. Length up to 25 cm. Many have robust, thick shells that are dark and can be tinged with yellow or green. *Reproduction:* Very intricate process. The larvae, called glochidea, develop inside the female and are released into the water, where they attach to a host fish. The larva parasitizes the fish before dropping to the stream bottom to develop. *Food:* Filter feeder. Filters plankton and organic debris from the water. Fed upon by fish, ducks, mammals. *Pollution Tolerance:* Moderately intolerant.

Image source: TVA, Water Management, Clean Water Initiative

RIGHT-HAND OR GILLED SNAIL
(Class Gastropoda, Family Lymnaeidae)

Description: Length up to 25 mm. Swirling shell opening right side (with shell pointing up and opening facing viewer). Has gills for breathing. Color brown, gray or black. May be covered with algae. *Reproduction:* Gelatinous egg mass deposited on rocks and other debris. *Food:* Algae, plant and animal debris. Fed upon by many fish, birds, and turtles. *Pollution Tolerance:* Fairly tolerant.

Image source: TVA, Water Management, Clean Water Initiative

SCUD or SIDESWIMMER
(Phylum Arthropoda, Class Crustacea, Order Amphipoda, Family Grammaridae and Talitridae)

Description: Length to 20 mm. humpbacked, extremely flattened side to side. Looks like a flea. Color varies, usually gray. Antennae visible. Swims on side or back. *Reproduction:* female holds eggs until they hatch. *Food:* Scavenge plant and animal debris. Food for many aquatic animals. *Pollution Tolerance:* Fairly tolerant.

Image source: TVA, Water Management, Clean Water Initiative

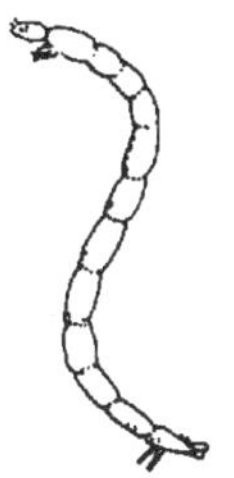

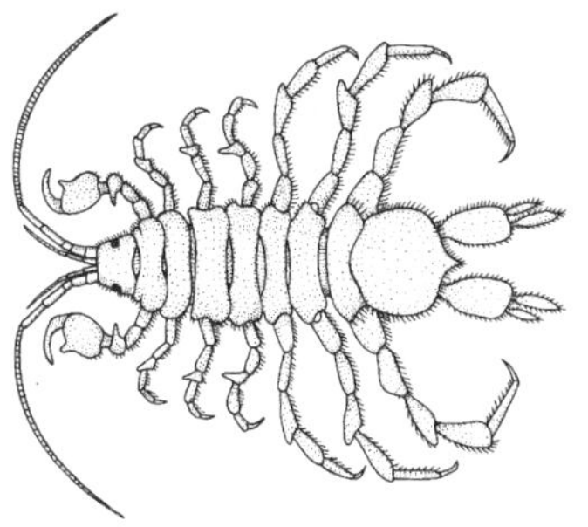

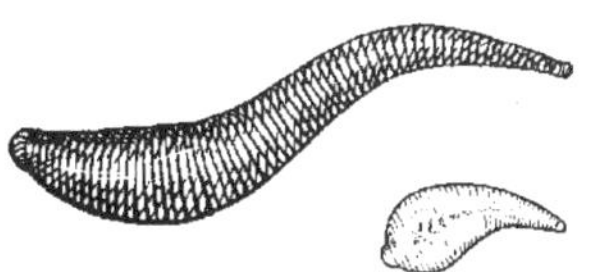

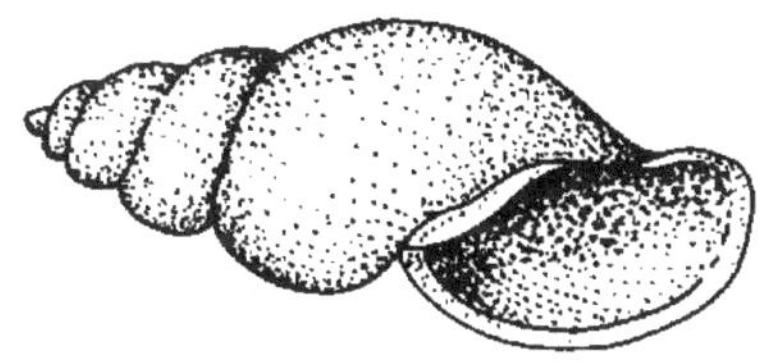

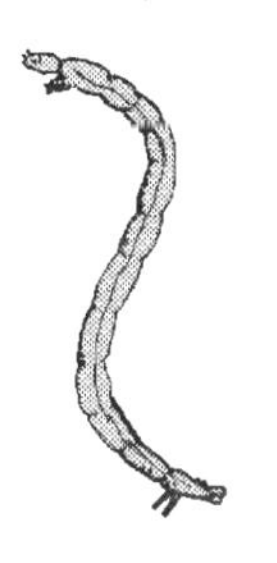

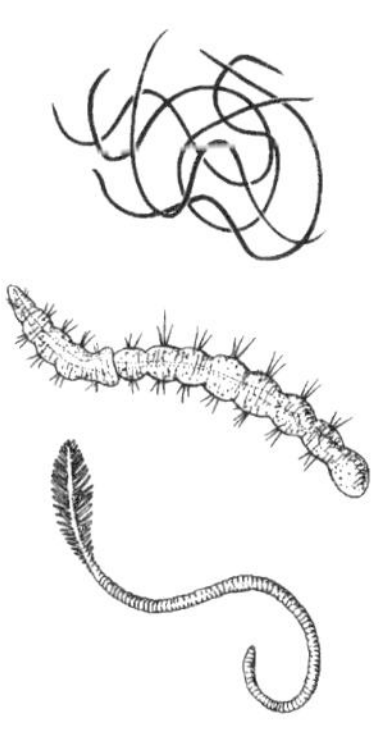

AQUATIC SOWBUG

(Phylum Arthropoda, Class Crustacea, Order Isopoda, Family Asellidae)

Description: Length up to 25 mm. Somewhat flattened top to bottom; resembles terrestrial version. Seven pairs of legs. Color varies but usually gray. *Reproduction:* Female carries the eggs carried under her abdomen until they hatch. *Food:* Scavenge both plant and animal debris. Food for other animals. *Pollution Tolerance:* Fairly tolerant.

Image source: TVA, Water Management, Clean Water Initiative

MIDGE

(Non-biting) (Order Diptera, Family Chironomidae)

Description: Adult: Length to 10 mm. Resembles a mosquito but thinner body parts. Swarms to light at night. Fuzzy antennae in males. Larva: Length to 15 mm. Elongated, cylindrical, slender, wormlike, wriggle rapidly out of the water. Color gold, brown, green, tan to black. *Reproduction:* Female deposits gelatinous egg mass on water surface or submerged vegetation. *Food:* Algae and organic plant debris. Some feed on insect larvae. Food for many organisms. *Pollution Tolerance:* Fairly tolerant.

Image source: TVA, Water Management, Clean Water Initiative

LEFT-HAND OR POUCH SNAIL

(Class Gastropoda, Family Physidae)

Description: Length up to 12 mm. Shell opening is on the left side (with shell pointing up and opening facing viewer). No gills but a sac-like lung. Colors are brown, gray, or black. May be algae-colored. *Reproduction:* Gelatinous egg mass deposited on rocks or other debris. *Food:* Algae, plant and animal debris. Fed upon by many fish, birds, and turtles. *Pollution Tolerance:* Very tolerant.

Image source: TVA, Water Management, Clean Water Initiative

LEECH

(Phylum Annelida, Class Hirudinea)

Description: Length to 70 mm. Worm-like, flattened lengthwise, with a sucker at each end. Color green, black, brown, or gray. Some have bright yellow or red patterns. *Reproduction:* Lay eggs in the water. Eggs are very hardy and can dry out. *Food:* Some species feed on blood, others eat detritus—decaying plant or animal debris. Not important as a food source for aquatic animals. *Pollution Tolerance:* Very tolerant.

Image source: TVA, Water Management, Clean Water Initiative

AQUATIC WORM

(Phylum Annelida, Class Oligochaeta, Family Tubificidae and Lumbricidae)

Description: Length up to 30 mm. Slender aquatic earthworm, segmented. Color reddish brown or gray. *Reproduction:* Hermaphroditic as in earthworms. Fertilization and development occur in a cocoon. *Food:* Ingest mud and filter out organic debris. Fed upon by bottom dwellers. Bloodworms or Tubifex sp. are good indicators of pollution. *Pollution Tolerance:* Very tolerant.

Image source: TVA, Water Management, Clean Water Initiative

BLOODWORM MIDGE

(Order Diptera, Family Chironomidae)

Description: Adult: Length to 10 mm. Resembles a mosquito but body parts thinner. Swarm to light at night. Do not bite. Larva: Length to 30 mm. Elongated, cylindrical, slender. Blood red in color. *Reproduction:* Female deposits gelatinous mass of eggs on surface of water or on submerged vegetation. *Food:* Algae and organic plant debris. Food for many organisms. *Pollution Tolerance:* Very tolerant.

Image source: TVA, Water Management, Clean Water Initiative

Transparency Master

TEACHER TABLE 3-1				
Model of Water-Quality Index				
Test	**Test Results Mean Values (%)**	**Q-Value**	**Weighting Factor**	**Total**
1. Dissolved oxygen	46% Sat	38	0.17	
2. Fecal coliform	16,000 colonies/ 100 mL	8	0.16	
3. pH	7.7 units	91	0.11	
4. BOD	4.89 mg/L	63	0.11	
5. Temperature change	$T_{site}1$ ___°C $T_{site}2$ ___°C Δ T = _01°C	93	0.10	
6. Phosphate	0.63 mg/L	51	0.10	
7. Nitrate	10.37 mg/L	48	0.10	
8. Turbidity	0.30 meters	5	0.08	
9. Total solids	475 mg/L	36	0.07	
	Overall Water-Quality Index			______ %

Overall Water-Quality Index	**Quality of Water**
90–100%	Excellent
70–90%	Good
50–70%	Medium
25–50%	Bad
0–25%	Very Bad

Name

Creating and Using an Index

As judges at an Olympic figure-skating competition evaluate each skater's performance, they consider many factors. Tough jumps count more than easy ones. Certain types of errors may cost a performer a chance at a medal, while others do not. This evaluation scheme is a type of index. An **index** is a rating system that assigns a value to an object or process, or to specific qualities it may possess.

People use an index to get a quick picture of the quality or status of an item. For instance, index systems have been developed to help people buy cars. These systems report and rank conditions, services, and problems that automobile purchasers should consider. An index that involves calculating human choice and individual differences is less likely to provide a single overall value than an index based on quantitative measures, such as the number of organisms in a water sample.

Even though the idea of employing an index may not have occurred to you, you probably encounter them quite frequently. Grades are **indices** (IHN duh seez) (the plural of index) of academic achievement. So, too, are movie guides, Neilson television ratings, wind chill factors, pollen counts, and product ratings in *Consumer Reports*.

In order to analyze, share, and compare results among one another, people who use an index must develop a uniform procedure (**protocol**) for its application and significance. These protocols provide the criteria by which information—in our case, river and stream quality—are measured.

Though identifying individual organisms and structures in a stream may be relatively easy, seeing the overall picture is more difficult, unless the work produces a quantitative picture of the stream. The indices used in studying a river or stream offer a mathematical picture that reduces many values to one or two overall numbers.

For instance, when a group of water experts for the National Sanitation Foundation wanted to compare water quality anywhere in the country, they developed the **overall Water Quality Index** (WQI). This index, described later in this lesson (and fully explored in *Rivers Chemistry*), uses raw data, weighting factors, and graphs to formulate a total score and water quality rating for a body of water.

A **weighting factor** is the value or points given to a factor in an index to make it count for more in a quantitative analysis. The most important factor receives the most points, and the least receives the fewest points (or vice versa). In terms of the Water Quality Index, having the most points is the best.

Questions

1. What kind of index or indices would you think should be used for evaluating a river or stream and its environment? Describe what factors they might measure.
2. Are indices neutral? Give examples to support your point of view.
3. Rating systems or indices are sometimes developed by panels of experts in a particular field; then everyone accepts and uses the index. Identify one of those indices and determine how it was developed and who uses it.
4. Are there limits to what people can evaluate meaningfully using an index system? Explain your reasoning and offer examples.

Biological Indices of Water Quality

In *Rivers Biology,* you will determine the quality of your local river or stream by investigating and calculating several different types of indices of water quality. Just as the doctor performs many observations and runs different tests to determine the health of a patient, so the scientist that studies the river must have more than one test to assess its health. Multiple indices provide a broader picture of the river's health. Moreover, when two tests measure some of the same parameters, they tend to do so in different ways, which allows for cross-checking and problem solving.

Why Assess Biological Water Quality?

Use of an index allows you to observe and quantify fluctuations in river or stream water quality. Using an index ratio over an extended period of time can indicate whether the water is becoming more polluted or cleaner. Assessments of biologically-based stream quality play important roles in local, state, and federal governmental decision making. With indices, organizations can rank rivers to decide which most need remedial action, which most deserve the limited resources available for restoration or maintenance, and what types of resources are needed.

Indexed assessments also provide baselines for evaluating proposed river work or development. Such information helps professionals gauge the effect of particular changes to the river watershed, particularly those caused by human development. Even after removal of a source of pollution, full recovery of a natural community may require considerable time. Therefore, scientists must assess the health of a stream on a recurring basis and compare the results with other similar streams.

Biological Indices

Keeping a stream clean and safe—for you and for the animals and plants living there—requires studying it and identifying the relationships between its water quality and its inhabitants. To really know a river, investigation must include evaluation of the aquatic life of the river's biological (**biotic**) relationships. The U.S. Environmental Protection Agency (EPA) recommends that such investigations gather as much information as possible on the following types of organisms:

- fish
- plankton

- **macroinvertebrates** (MAK roh ihn VERR teh brayts). **Invertebrate** (without backbone) organisms visible without the aid of a microscope.
- **microinvertebrates** (MY kroh ihn VERR teh brayts). Invertebrate organisms so small that seeing them requires a microscope.
- **periphyton** (PEHR ih fy tuhn). Organisms attached to underwater surfaces such as sponges and rotifers.
- **macrophytes** (MAK roh fyts). Larger plant organisms.

Though most surveys, including those in *Rivers Biology,* cannot realistically gather data on all the categories, surveying several parameters is better than doing just one. For instance, many agencies monitor fish populations; fish are easy to identify, even in the smallest streams. Fish are highly mobile, however, while macroinvertebrates are less mobile, meaning that they are more responsive to localized conditions. So a macroinvertebrate study often yields valuable information a fish survey may not.

In *Rivers Biology,* most of the methods you will use for determining water quality involve data on the biological characteristics of your river or stream. To gather such data, you will do field collection and identification of specific organisms. When you have completed your observations, you will have a large set of data. As you recall, scientists use an index to summarize their observations in a simple way and to express their results with a single value.

Using one index, as an indicator of water quality you will assess the presence of organisms of various tolerance to pollution. (In terms of biology, **tolerance** refers to an organism's ability to exist in an environment affected by the relevant factor.) For another index, you'll statistically analyze your observations of living organisms to measure the ability of that aquatic site to sustain a diversity of life. Later in this lesson, you will also learn how a set of chemical tests of river or stream water also indicates water quality. The ability of an organism to survive is influenced in large measure by the chemistry of its environment. Natural forces affect the chemical status of the water; today, however, many chemical impediments to aquatic life result from human activity.

In this lesson, you will learn about such indices useful in assessing the biological health of rivers and streams. In the next lesson, you will learn the practical biological sampling techniques to use when gathering data for such indices. In Lesson 4, you will gather such data and calculate these biological water-quality indices for your local river or stream site. The field techniques and calculations for some chemical tests related to water quality will be developed later in *Rivers Biology.*

DRIFTWOOD

In 1992, 43 states reported 930 pollution-caused fish kills affecting more than 5 million fish.

Benthic Macroinvertebrate Pollution-Tolerance Index

Two freshwater systems indices used in *Rivers Biology* are based on biological observations of species presence. The first is the benthic macroinvertebrate pollution-tolerance index. **Benthic macroinvertebrates** (BEN thik MAK roh ihn VERR teh brayts) are bottom-dwelling invertebrate organisms visible without the aid of a microscope. Compared with fish or turtles, benthic macroinvertebrates are relatively immobile. Because they remain in one general area of the stream and adapt to that environment, they are an important means of assessing the aquatic environment. Their survival depends on water temperature, food availability, amount of oxygen in the water, flow velocity, and type of bottom surface of the streambed. Every river and stream has benthic macroinvertebrates at some time during the year unless it is very, very polluted.

Different types of benthic macroinvertebrates have specific limits of pollution tolerance, making them good indicator organisms for pollution. Some benthic macroinvertebrates can survive only in favorable conditions; they are intolerant to pollution or habitat alteration. Many of these species have multiple-year life cycles and are restricted to their immediate environment because they do not move at all. Their absence provides an early warning signal of water-quality degradation. Others can survive in a range that includes conditions unfavorable to all but a few species; these are called "pollution tolerant" organisms.

The **benthic macroinvertebrate pollution-tolerance index** places benthic macroinvertebrates in groups according to their level of pollution tolerance. Each listing is a taxonomic group. Each **taxon** (**taxa** is the plural) is a category into which related organisms are placed, such as apparent species. When analyzing survey results, taxa in the most pollution-tolerant group are weighted more heavily, with the least tolerant weighted most lightly. Because the benthic macroinvertebrate indexing system is relatively quick to set up and perform, it provides a rapid means of assessing stream quality.

The Diversity Index

The second biological assessment in *Rivers Biology* is a **diversity index,** which represents the number of species or types of organisms present in a biological community. In most streams, having many species is preferable. High diversity indicates that conditions support a wide variety of aquatic life. Natural biotic communities unaffected by humans are typically characterized by the presence of a few species with many individuals or many species with a few individuals.

Pollution reduces the number of species present in a community. So a high diversity index is an indicator of good balance between stresses and natural energy flow. Polluted waters generally have lower diversity-index values.

In special circumstances, however, very clean water may have low diversity. For instance, cold temperatures may reduce the number of species that can survive. Seasonal variations in diversity can also affect water temperature and, thus, diversity. For instance, as the temperature of the water rises in late summer, species found in colder water become dormant, so they may not be captured in surveys.

The diversity index used in *Rivers Biology* is based on the diversity indexing system developed by scientists C. E. Shannon and W. Wiener. Their index incorporates not only the number of species present but the proportions of the entire community each species represents. This system derives from the premise that a community with relatively balanced species populations is more stable than a community in which a single species comprises a high percentage of the population. If this single species were to become threatened, the entire community would be in danger of collapse.

In many cases, the diversity index analysis uses the same data gathered for the benthic macroinvertebrate tolerance index. The diversity index involves a bit more complexity than that benthic index, however. It involves the additional step of calculating the proportion of the types of organisms found in the sample.

When comparing the conservation value of various locations, a site with high diversity would seem to be more valuable than a similar one with low diversity. In general, this is probably true, but in some cases one must consider other variables. For example, a spring or pothole may have a very poor aquatic species diversity, but the uniqueness of the inhabitants and the extremes of the habitat that lead to the low diversity may also need consideration. Such environments may be rare, even endangered, indicating the need for human protection of the rare organisms living here.

Limiting Factors Affecting Aquatic Samples

Though pollution influences the diversity and density of an aquatic community, it is not the only factor affecting a particular sample and the resulting index. Scientists who want the most accurate results from collecting and analyzing samples and constructing biological indices consider limiting factors. A **limiting factor** is a variable, either natural or human made, that limits conditions or actions.

Whenever you collect a biological sample from a natural setting, note as many as possible of the factors discussed in the following passages. If possible, collect samples from a site at more than one time. When collecting samples, be sure to carefully follow the collection procedures described in the next lesson.

- Season of year. Spring floods wash away some organisms and introduce others into river or stream sections that may not yet have any established populations of those organisms. Autumn drought can relocate or kill populations of organisms. Winter ice and snow can affect life cycles. Changes in water temperature through the season are a very important criterion for aquatic life.

- Sampling location. Water biologists and river watchers need to know the habitats of the organisms they are observing. For instance, the ideal site for pollution-intolerant benthic macroinvertebrates is a **riffle,** a relatively shallow area of the stream over which water flows unevenly. (Some, however, are adapted to living along the bank in slower water.)
- Life cycles. Many insect populations grow actively in some seasons of the year and are present only as eggs or pupae in other seasons. Some aquatic insects have larval forms that emerge as adults mainly in the springtime.
- Type of substrate. The **substrate** (SUB strayt) is the material at the bottom of the streambed, on which organisms live or reproduce. Rocky bottoms provide more habitats than silty or sandy bottoms. The energy of rapidly moving water can change a silty or sandy bottom—thus eliminating habitats—much more readily than it can change a rocky bottom.
- Food sources. Organisms are dependent on food availability. Food from plant material is more abundant in river and stream sections bordered by forested areas.
- Stream velocity. Velocity, also known as **flow rate,** is the speed with which the water is moving. The optimal speed of a stream or river for supporting organisms is 15–60 cm per second (0.5–2.0 feet per second). In faster streams, only specially-adapted organisms can survive. In slower streams, the water may drop silt, which may cover the organisms, habitat, breeding areas, or prey. High silt volumes can bring organic materials that decay, reducing oxygen in the water. Compared with faster water, slower water has less oxygen, which animals need.
- Sampling techniques. Incorrect mesh size of the netting or screen, or improper use of equipment, may allow small organisms to escape or go unobserved. This would decrease apparent diversity in the sample.

Driftwood

"Rivers and all creatures that inhabit the water were put here for wise men to contemplate and fools to ignore."

Don Orth
Blacksburg, VA

Questions

1. Explain what an index is designed to accomplish.
2. Explain why an aquatic ecosystem with low diversity might be unstable or considered unhealthy.
3. How does a diversity index differ from a macroinvertebrate pollution tolerance index?
4. Identify three factors that affect the diversity of a community and, in your own words, explain their effect.

Name

Benthic Macroinvertebrates and Pollution Tolerance

As indicators for water quality, scientists have placed each benthic macroinvertebrate species into one of four groups, according to its tolerance for pollution. The groups are: most intolerant; moderately intolerant; fairly tolerant; and most tolerant. Scientists, in preparing the index results, chose that taxa in the most pollution-tolerant group are weighted more heavily, with the least tolerant weighted most lightly. The greater the number of specimens from a group in a sample of river or stream water, the more likely that water is to have the pollution level that corresponds to that group.

For example, a stream with many intolerant organisms has little pollution. If that stream joins another stream downstream and a survey below that junction shows only moderately tolerant organisms, the water quality has become worse. Other work can then investigate what types of degradation the water has experienced and why. The presence of particular macroinvertebrates may indicate something is happening, but it does not reveal causes.

The pollution-tolerance groups presented in *Rivers Biology* are very general. Most families of benthic macroinvertebrates have members in more than one pollution-tolerance group. Within families, different species may have different levels of pollution tolerance depending on the climate and other regional variables. Not every species in a taxonomic family, not even every organism within a genus, has the same level of pollution tolerance. Pollution-tolerant organisms like clean water as well as intolerant organism do, but they can also survive in polluted places, while those others cannot.

Group I: Pollution-Intolerant Organisms

Benthic macroinvertebrates in the pollution-intolerant group cannot survive in habitats affected by nutrient pollution such as sewage or fertilizer. They are especially intolerant of low levels of dissolved oxygen. If the pollution increases in a stream or river, all organisms from Group I will die out. These organisms are also referred to as pollution-sensitive.

These clean-water macroinvertebrates are generally diverse and relatively abundant. In environments in which they thrive, many different types of organisms are present, and, except for seasonal appearances, none dominate the community by number. Many members of this group feed on decaying leaf matter in the stream. Some are predators and depend on naturally available food sources.

DRIFTWOOD

"If you could tomorrow morning make water clean in the world, you would have done, in one fell swoop, the best thing you could have done for improving human health by improving environmental quality."

William C. Clark, Racine, WI, April 1988

Examples of common Group-I benthic macroinvertebrates are shown in Figure 3-1. Cold, clear streams with high oxygen levels typically have significant populations of these organisms. Stoneflies cannot tolerate any type of pollution, so they are sometimes called the canary of clean water. Alderflies and dobsonflies, which also belong in Group I, are some of the largest aquatic insects, found in a variety of habitats along the stream bottom and riffle areas. Of Group I taxa, snipeflies are the most tolerant.

Group II: Moderately Intolerant Organisms

Organisms in Group II are more tolerant of nutrient-enrichment from agricultural fertilizer, municipal sewage, and other sources. As dissolved oxygen levels drop, pollution-intolerant forms decline in number and may even disappear. Some natural pollution actually increases the number of these fairly pollution-intolerant organisms. Such forces may include seasonal stagnation and heat in the late summer.

In many parts of the country such as the Midwest, finding a stream that is not slightly enriched by phosphates and nitrates from agricultural runoff is almost impossible. That enrichment affects the types of organisms found in the water. Sedimentation also lowers the water quality of a stream by suffocating organisms, thus lowering the tolerance value. Sedimentation from land development and agriculture is a major cause of water-quality deterioration.

Most mayfly nymphs are fairly tolerant of lower oxygen levels, and caddisflies are even more tolerant. These two groups make up the greatest numbers of organisms found in Group II. Riffle beetles and water penny beetles flourish when warmer temperatures and decreased oxygen support find the increased growth of algae, which they consume. Cranefly larva and crayfish find a home in leaf packs, riffles, and under stones or debris. Both can stand some pollution. Dragonfly and damselfly nymphs are found along the margins of streams or in pooling water, attached to vegetation. As predators, they are affected by toxic materials passed up through the food chain through their prey. Once mussels become adults, they hardly move. Fingernail clams, which once covered the bottoms of many rivers and served as important waterfowl food, have been greatly reduced by pollution and sedimentation. Examples of common Group II benthic macroinvertebrates are shown in Figure 3-1.

Group III: Fairly Tolerant Organisms

Group III contains macroinvertebrates that can accept low-oxygen conditions and fairly high levels of nutrient enrichment. These tolerant organisms tend to be either scavengers or omnivores and are bottom dwellers. In more polluted environments, they replace the less-tolerant organisms.

The blackfly larva is often quite abundant below sewage treatment plants. Sowbugs and scuds are **detritus** eaters, meaning they live mainly on decomposing organic matter. (**Organic matter** is anything that is or was alive.) Some midge larvae feed on a variety of food sources and are generally tolerant of pollution. The gill-breathing or right-hand snail also feeds on a variety of plant and animal matter attached to plants, branches, or rocks. Examples of common Group III benthic macroinvertebrates are shown in Figure 3-1.

Group IV: Pollution-Tolerant Organisms

Group IV organisms are very tolerant to low levels of oxygen and to severe nutrient pollution. Most polluted streams contain huge numbers of these organisms. Human-made food sources, such as agricultural runoff and sewage, provide rich nutrient sources to support those large populations. Pollution-tolerant organisms are usually associated with bottom sediments. Some have special adaptations to obtaining oxygen from the surface, because so little is present in the water.

Aquatic worms are capable of living almost without oxygen. Leeches are adapted to very stressed environments and are omnivores. The adaptive ability of air-breathing or pouched (left-hand) snails to obtain oxygen allows them to live in almost any water. Bloodworms, a type of midge larvae, have a hemoglobin-type blood that assists them in moving oxygen through the body. Examples of common Group IV benthic macroinvertebrates are shown in Figure 3-1.

Questions

1. Describe an environment in which you are likely to find Group I benthic macroinvertebrates.
2. What factors in a relatively polluted environment may actually support pollution-tolerant species?
3. In what portion of the stream or river environment do most benthic macroinvertebrates live?
4. Predict the type (or types) of benthic macroinvertebrates you are likely to find in your local river or stream.
5. Summarize the conditions and the animals living in the river or stream for each tolerance group.

GROUP 1 – These organisms are considered intolerant of pollution

Stonefly Nymph

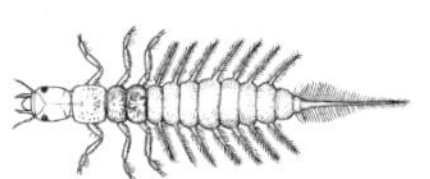

Alderfly Larva

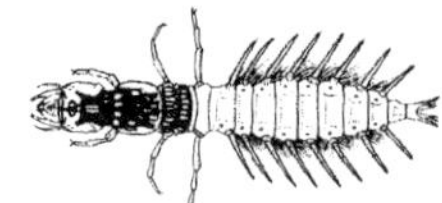

Dobsonfly Larva

Snipefly Larva

GROUP 2 – These organisms are considered moderately intolerant of pollution

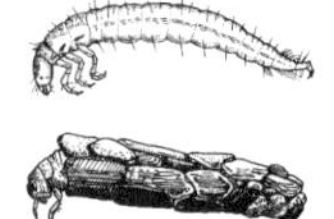

Caddisfly Larva

Mayfly Nymph

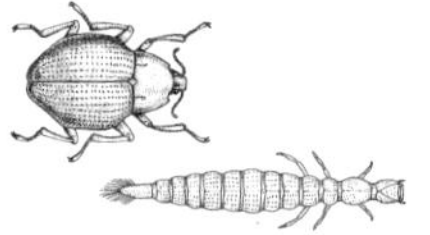

Adult Larva Rifle Beetle

Water Penny Larvae

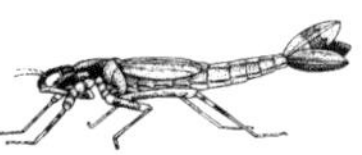

Damselfly Nymph

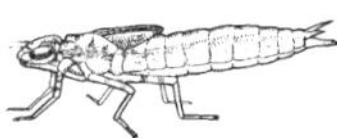

Dragonfly Nymph

Cranefly Larva

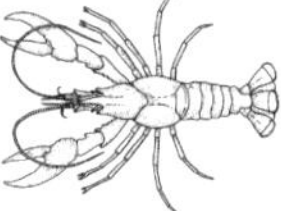

Crayfish

Clam/Mussel

GROUP 3 – These organisms are considered fairly tolerant of pollution

Black Fly Larva

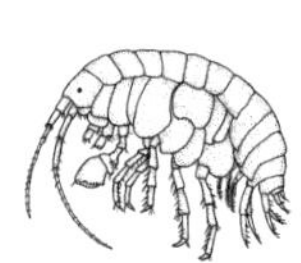

Scud

Right-Hand/ Other Snail

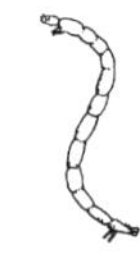

Midge Larva

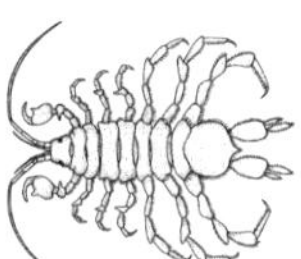

Sowbug

GROUP 4 – These organisms are considered very tolerant of pollution

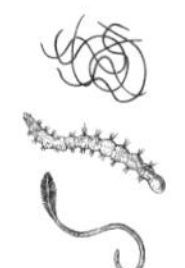

Aquatic Worm

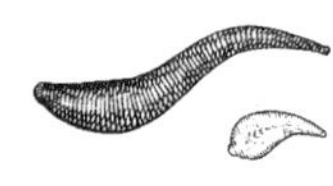

Leech

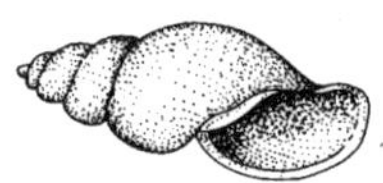

Pouch/Left-Hand Snail

Bloodworm Midge Larva

Source of Images: TVA, Water Management, Clean Water Initiative

Figure 3-1: Benthic Macroinvertebrate Pollution-Tolerance and Field Identification Reference Sheet

Name

STUDENT

ACTIVITY 3.4

The Water-Quality Index

Purpose

To learn how to compute a water-quality index based on chemical tests and to be able to describe water quality using index values.

Background

By performing and analyzing **abiotic** (not biological) tests on river or stream water, water biologists and river watchers can compare water-quality values based on observations of living organisms with those based on chemical parameters. The National Sanitation Foundation (NSF) uses nine tests to determine water quality: dissolved oxygen, fecal coliform, pH, biochemical oxygen demand, temperature change, phosphates, nitrates, turbidity, and total solids.

In *Rivers Biology,* you will learn the definitions, significance, and procedures for some of these tests, specifically dissolved oxygen, biochemical oxygen demand, and fecal coliform, because those are the factors that most directly affect the health of organisms living in the river and humans using the river. After performing these tests on your local river or stream, you will use your results to develop a final index value, the overall Water Quality Index (WQI). Ideally, your class, or a class doing *Rivers Chemistry* on the same river or stream site, will conduct all nine chemical tests and use that data to further collaborate biological studies. (*Rivers Chemistry* covers all the tests.)

Conducting each of these nine water-quality tests gives investigators data, but data for different tests is in different units. For instance, the unit for fecal coliform is colonies/100 mL, but the unit for temperature is degrees Celsius. The unit for each type of test is shown in Table 3-1. So how do investigators add up the results of all these tests if the results are in different units?

If you add 5 oranges and 6 apples, you do not get 11 oranges, nor do you get 11 apples. You get 11 fruits. In other words, you must convert the 5 oranges and 6 apples into a unit that is common for both; in this case, the common unit is fruit. Water-quality experts have developed a unit that is common to all nine tests. It is called **Q-value percent.** Determining overall water quality or comparing the results of different types of tests requires converting results from each of the nine tests to the common Q-value. Each test for water quality has its own Q-value chart that facilitates this conversion. You will find and use these graphs in lessons as you progress through *Rivers Biology*. An example of such a graph is Figure 3-2.

Here's how to do such a conversion. Assume you analyzed a sample of water and found 300 mg/L of total solids. Now you need to convert total solids from mg/L to Q-value percent. Find 300 mg/L on the horizontal axis of Figure 3-2. The 300 mg/L line intersects the curve at 60 percent. Therefore, the Q-value for 300 mg/L of total solids is 60 percent.

Just as exams count more than quizzes, some of the nine water-quality tests have a higher weighting factor than others. Table 3-1

Materials

Per group

- calculator

For the teacher

- overhead transparency of Model of Water-Quality Index, Teacher Table 3-1
- overhead projector

lists the nine tests and their individual weighting factors. Dissolved oxygen has the highest weighting factor of 0.17, while total solids has the lowest, 0.07. The weighting factor is actually the decimal equivalent of its percent. For example, dissolved oxygen's weighting factor of 0.17 is equivalent to 17 percent of the total grade; total solids count 7 percent of the final grade. If you add the weighting factors for all nine water-quality tests, they equal 1.00, or 100 percent. Try it, using Table 3-1.

TABLE 3-1

Units and Weighting Factors for the Nine Water-Quality Tests

Test	Units	Weighting Factor	Percent	Total (%)
Dissolved Oxygen	% Sat	0.17	= 17%	
Fecal Coliform	colonies/100mL	0.16	= 16%	
pH	units	0.11	= 11%	
BOD	mg/L	0.11	= 11%	
Temperature Change	°C	0.10	= 10%	
Phosphate	mg/L	0.10	= 10%	
Nitrate	mg/L	0.10	= 10%	
Turbidity	meters/feet or JTU	0.08	= 08%	
Total Solids	mg/L	0.07	= 07%	

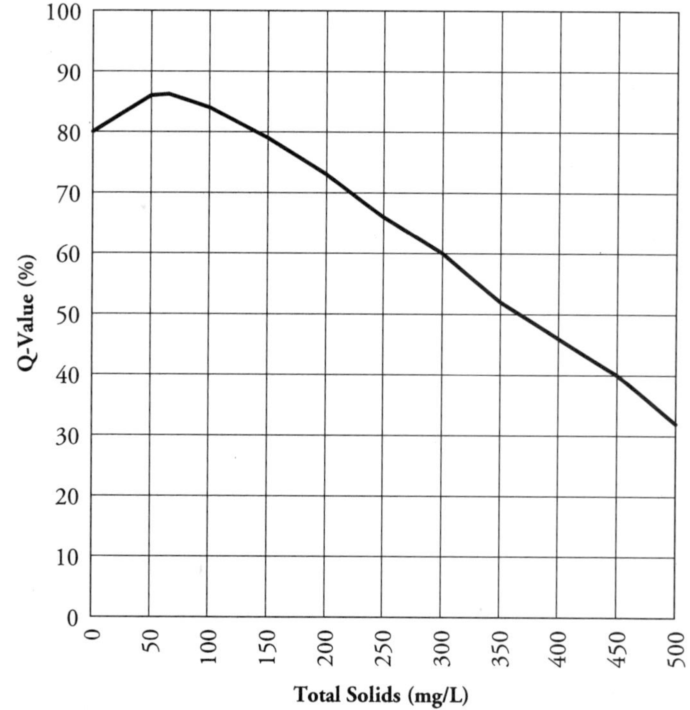

Figure 3-2: Graph for Converting Total Solids from mg/L to Q-value

Procedure

1. When a sample of river water was analyzed, the information in the table following was obtained. (The Q-values have already been calculated.)

Test	Test Results (Mean Values)	Q-Value	Weighting Factor	Total (%)
1. Dissolved Oxygen	46% Sat	38	0.17	
2. Fecal Coliform	16,000 colonies/100mL	8	0.16	
3. pH	7.7 units	91	0.11	
4. BOD	4.89 mg/L	63	0.11	
5. Temperature Change	Δ T = 01°C	93	0.10	
6. Phosphate	0.63 mg/L	51	0.10	
7. Nitrate	10.37 mg/L	48	0.10	
8. Turbidity	0.30 meters/feet or JTU	5	0.08	
9. Total Solids	475 mg/L	36	0.07	

OVERALL WATER-QUALITY INDEX _____ %

2. For each test, determine the total percentage by multiplying the Q-value by its corresponding weighting factor. This will give you a decimal score (or percent, if you use the percent values instead of the decimal values). Record your answers in the blank column. (Make your own table if your teacher requests this.)
3. Add all the scores (or percent values) for all the tests to obtain the overall water-quality index (WQI).
4. (If you do not have data for all tests, which will be the case with your work in *Rivers Biology,* you will sum the weights of all the tests for which you do have data. Then divide that total by the sum of the weighting factors for those tests. For example, if the total calculated water quality with only three tests was 35% and the total of the weighting factors for those three tests was 0.44, then the water-quality index is 35% divided by 0.44 = 79%.)

Overall Water-Quality Index	Quality of Water
90–100%	Excellent
70–89%	Good
50–69%	Medium
25–49%	Bad
0–24%	Very Bad

Analyses and Conclusions

1. Using your water-quality index, and the preceding chart, rate this river or stream water from excellent to very bad.
2. Why is the Q-value used?
3. How is the WQI obtained?

Critical Thinking Questions

1. Which chemical test do you think would be most important to perform on water that people might swim in? Why?
2. Which chemical test do you think most affects environmental quality for aquatic animals? Why?
3. Do you think that chemical water-quality results might support or conflict with biological water-quality test results? Explain your reasoning.

Keeping Your Journal

1. List several other scientific fields besides chemistry that could be helpful to biologists studying riverine environments, and explain how they would be helpful.
2. Imagine you are an aquatic animal living in your local river or stream. Describe how several of these chemical variables affect your life.

TEACHER NOTES

LESSON 4 Biological Surveys

Focus

Students will observe the local river or stream and its habitats in an organized manner and determine the types and number of organisms found in one or more biological surveys. They will then analyze their data to construct several indices of the biological health of the local river or stream.

Learner Outcomes

Students will:

1. Learn to accurately collect benthic macroinvertebrate samples from a river or stream.
2. Learn the components of a riverine habitat assessment.
3. Identify benthic macroinvertebrates from samples collected at the river or stream.
4. Learn to use the organisms collected to describe the diversity of the river or stream.
5. Learn to use the data collected to assign a value to the quality of the local river or stream based on biotic indicators.
6. Complete a habitat assessment of the local river or stream that contributes to a determination of water quality.

Time

Five to six class periods of 40–50 minutes per period, plus a field trip to a local river or stream

DAY 1: Student Information 4.1: Looking at the River

DAY 2: Student Information 4.2: Guide to Sampling Benthic Macroinvertebrates

DAY 3: Student Information 4.3: Riverine Habitats

FIELD TRIP: Student Activity 4.4: Determining the Benthic Macroinvertebrate Index for Your River or Stream
Student Activity 4.5: Completing a Riverine Habitat Survey

DAY 4: Complete Student Activity 4.4: Determining the Benthic Macroinvertebrate Index for Your River or Stream
Student Activity 4.6: Determining the Diversity Index for Your River or Stream

DAY 5: Student Assessment 4.7: A River Diorama (complete as homework)

Advance Preparation

If possible, do more than one field trip during *Rivers Biology*. In most cases, students achieve best results if a first field trip focuses just on the biological surveys in Lesson 4 and a subsequent field trip focuses on chemical tests for Lessons 5 and 6. If you must do only one field trip, do the field-trip activities in Lesson 4 (Student Activities 4.4 through 4.6) after students have prepared in Lesson 5 to do Student Activity 5.4 and prepared in Lesson 6 to do Student

Activity 6.3. Then you can have students do all the field-trip activities for *Rivers Biology* on one field trip. (If you do multiple field trips, you can also have students repeat the biological surveys on later trips.)

Prepare overhead transparencies of Teacher Table 4-1 (Benthic Macroinvertebrates in Sample), Teacher Table 4-2 (Model of Benthic Macroinvertebrate Pollution-Tolerance Index), and Teacher Table 4-3 (Model of Diversity Index), located at the end of these Teacher Notes.

Prepare to supply students with Student Information, Activity, and Assessment sheets 4.1 through 4.7. If students will be preparing permanent collections of benthic macroinvertebrates during their field trip, prepare instructional materials as appropriate.

Gather necessary equipment and materials. For listings of appropriate identification guides for macroinvertebrates (and plants and animals), see Appendix C, Resources. Identification guides produced by your state make macroinvertebrate identification more specific and easier.

Laminate one copy per group of the Benthic Macroinvertebrate Pollution-Tolerance and Field Identification Reference Sheet (Figure 3-1 from Student Information 3.3). Laminate the topographic map of the field-study site. You may wish to laminate a materials list for each field-site activity.

The habitat assessment form in Student Activity 4.5 is primarily a qualitative assessment, with some emphasis on biological observations. *Rivers Earth Science* contains a different habitat assessment form, one that is quantitative but less biologically oriented. If desired, you may either substitute that assessment for this one, or have different groups do different surveys and then compare results. The assessment in *Rivers Earth Science* was developed by the Illinois RiverWatch Network.

If you want students to do flow-rate measurement as part of the habitat assessment, prepare to distribute the specific handout from *Rivers Chemistry* or *Rivers Earth Science* on this procedure. If you want students to measure channel width as part of the assessment, prepare to distribute the specific handout from *Rivers Earth Science* on this procedure. These procedures require additional materials. Alternatively, you can obtain this information from other Rivers Curriculum classes or provide an approximation of channel width and obtain flow rate information for that stream and date from water resource agencies.

If desired, prepare a short lecture on local habitats and what students might observe. Review information in the unit introduction on field trip management. Be sure to visit the field site beforehand, assess all potential hazards, and obtain necessary permissions. If possible, arrange for additional adult supervision during the field trip. Local or state agency representatives can provide information on likely local species and may be willing to work with your class on this field-site visit.

Safety and Waste Disposal

Follow all field safety procedures presented in Lesson 1. Be sure students wear safety equipment where appropriate.

If chemical tests from Student Activities 5.4 and 6.3 are performed on a field trip during this lesson, carefully follow the test kit manufacturer's procedures and precautions. A used gallon milk container properly labeled works well to collect wastes from chemical tests. Solid wastes should be placed in plastic garbage bags. Bring the liquid and garbage back to the classroom for dehydration.

Materials

Student Information 4.2: Guide to Sampling Benthic Macroinvertebrates

For the teacher

set of 15 benthic macroinvertebrate cards (from pages 45 through 50)
D-net or triangular net
kick net
animal identification guide

Optional

bottles for making collections

Student Activity 4.4: Determining the Benthic Macroinvertebrate Index for Your River or Stream

Per student

life jacket and tow line (for deeper water sites)
rubber waders or boots (for shallower-water sites)
protective gloves (if water may have contaminants)
clipboard
journal
Student Information: 4.2 Guide to Sampling Benthic Macroinvertebrates

Per group

kick net (for rocky-bottom stream or river)
D-net or triangular collecting net (for muddy-bottom stream or river)
shallow white pan
white sheet, oilcloth, or tarp
2–3 collecting buckets
rinsing bucket
forceps
eyedropper
spoon
2-cm (1-in.) paintbrush
small watercolor brushes
laminated Benthic Macroinvertebrate Pollution-Tolerance and Field Identification Reference Sheet (Figure 3-1 from Student Information 3.3)
macroinvertebrate identification guide
hand lens or dissecting microscope

Per class
topographic map or other material for determining site location
first-aid kit
bottled water for washing
soap and towels
large plastic bags for collecting waste
Optional
variety of collecting bottles and clean jars with lids (for permanent collections)
alcohol (70% ethyl) for preservation
For the teacher
overhead transparency of Teacher Table 4-1, Benthic Macroinvertebrates in Sample
overhead transparency of Teacher Table 4-2, Model of Benthic Macroinvertebrate Pollution-Tolerance Index

Student Activity 4.5: Completing a Riverine Habitat Survey
Per student
life jacket and tow line (for deeper-water sites)
rubber waders or boots (for shallower-water sites)
journal
Per group
Student Information 4.3: Riverine Habitats
laminated topographic map of study site, 7.5-minute or 15-minute series
USGS pamphlet, *Topographic Map Symbols*
measuring tape, 50–100 meters
Celsius thermometer, alcohol-filled with a metal jacket, or electronic; or temperature probe
25–50 cm of light cord
stopwatch
clipboard or firm writing surface
blank paper
completed data tables for Student Activity 4.4: Determining the Benthic Macroinvertebrate Index for Your River or Stream
Per class
first-aid kit
bottled water for washing
soap and towels
large plastic bags for collecting waste
Optional
other material for determining site location
camera, digital if possible
compass

Student Activity 4.6: Determining the Diversity Index for Your River or Stream
Per group
completed data tables from Student Activity 4.4: Determining the Benthic Macroinvertebrate Index for Your River or Stream
calculator with logarithmic function

For the teacher
overhead transparency of Teacher Table 4-3, Model of Diversity Index

Student Assessment 4.7: A River Diorama
Per group
pictures or photocopies of animals found in your river
pictures or photocopies of plants found in your river
markers or colored pencils
cardboard box
Per class
poster board
finger or tempera paint
brushes
miscellaneous art materials
matte knife or strong scissors
computer and printer

Vocabulary

aesthetic value
bank erosion
bedrock
boulder
channel
channel capacity
cobble
cultural factor
density
dip net
D-net
effluent
embeddedness
erosion
erosion potential
evenness
flow
gradient
gravel
habitat assessment
hydrologic modification
kick net
levee
lower bank
migration
naturalness
nonpoint source pollution
parameter
point-source discharge
pool
richness
run
siltation
stream bank
triangular net
turbidity
upper bank
vegetated
watershed

Background for the Teacher

The field study activities in this lesson serve as essential activities in river study for students. Students can best study the animals of the river and stream by being at the water. You may demonstrate collecting techniques in the classroom, though demonstrations are most effective when performed at the river or stream. The activities in this lesson are the heart of *Rivers Biology*. Enjoy them, and share your excitement for the field as you lead the students out on the water. Once students learn how to perform biological surveys and to identify the kinds of organisms present in a riverine environment, they can go anywhere and in a short time make some informal judgment of the water quality of a river or stream.

Introducing the Lesson

1. If not already done, distribute permission forms for the field trip(s).
2. Make sure students have returned the signed Safety Guidelines and Contract (Student Information 1.3).
3. On the blackboard, write the word *collecting*. Have the students brainstorm everything they know about collecting on the river or stream they will be testing. Make a web or list of all their responses to identify what the students know about collecting on the river. Then add information they need.
4. Summarize this list and add to it by providing an overview of the activities that the class will be involved in during the next week or two.
5. Have students read, discuss, and answer the questions for Student Information 4.1: Looking at the River. (Answers for student sheets are in Appendix B.)
6. From the picture cards used in Lesson 3, select macroinvertebrates students may find at their field site. Choose some from each tolerance level. Show these to the students, explaining that these are what they will look for and collect during their field-site visit. In the field, have students use a laminated copy of Figure 3-1 (in Student Information 3-3), because the cards will be clumsy. Have students compare the cards to Figure 3-1 so they see this is a working sheet especially for field identification.
7. Have students read and discuss Student Information 4.2: Guide to Sampling Benthic Macroinvertebrates. Discuss what stream type they can expect. Demonstrate the use of equipment mentioned in Student Information 4.2 or that students will use at the river or stream. Have students practice with the equipment. Have students answer questions for Student Information 4.2; discuss answers.
8. Have students read and discuss Student Information 4.3: Riverine Habitats. Refer to the definitions for any unknown or misunderstood words. You may also choose to distribute Student Activity 4.5: Completing a Riverine Habitat Survey, so the class can familiarize themselves with the habitat survey sheet and how to use it. If desired, present a short lecture on what students might expect at the local river or stream habitat.
9. Have the students read and discuss Student Activity 4.4: Determining the Benthic Macroinvertebrate Index for Your River or Stream. If desired, show them the overhead of Teacher Table 4-1, Benthic Macroinvertebrates in Sample (located at end of these Teacher Notes, as are other Teacher Figures). Have them use these data to practice determining the pollution tolerance index, following the procedure in Student Activity 4.4. As appropriate, use an overhead transparency of Teacher Table 4-2, Model of

Benthic Macroinvertebrate Pollution-Tolerance Index, to have the students check how they performed the calculations.

10. Have the students read and discuss Student Activity 4.6: Determining the Diversity Index for Your River or Stream. Have them use the results of their practice pollution-tolerance index to practice determining the diversity index, following the procedure in Student Activity 4.6. As appropriate, use an overhead transparency of Teacher Table 4-3, Model of Diversity Index, located at the end of these Teacher Notes, to have the students check how they performed the calculations. For specific teaching activities that support the mathematics steps in this activity, see *Rivers Mathematics*. The steps are easily taken, but you may be more comfortable having a mathematics teacher explain the reasoning behind this mathematics process.
11. Discuss how the two different indices can portray the same data. Have the students place their indices from Student Activity 3.4 and sample results for 4.4 and 4.6 side by side and compare the final values. The values given in all three indices represent the same site. Indicate to students that when the results from all these indices indicate approximately the same water quality, the condition of the water is fairly consistent for both chemical inflow and other conditions that help to maintain the animal population.
12. Have students consider what could be happening if the water chemistry showed excellent water quality but the diversity and water tolerance indices were poor, or other variations on differing results.

Developing the Lesson

1. As mentioned in Student Information 4.3, in order to make field and lab data useful to others, students will need to know how to pinpoint exact site location, either using a compass and a topographic map, or through determining legal description. For specific student handouts on using maps and determining precise site location, see *Rivers Geography* and *Rivers Earth Science*.
2. Make sure all students have returned the permission forms, signed by a parent or guardian.
3. Tell students which surveys and tests they will perform at the field site during the upcoming field trip. Assign groups to each survey or test, or have the entire class do all activities.
4. Review field safety. Make sure students are familiar with the procedures, materials, and equipment for their field-trip assignments.
5. If students will be preparing permanent collections of benthic macroinvertebrates during their field trip, demonstrate the procedure you want them to use.
6. If you are going to conduct tests of dissolved oxygen, biochemical oxygen demand, or fecal coliform during this field trip, complete the preparatory student sheets in Lessons 5 and 6 before going on the field trip. If your students will be performing other tests of water quality, make sure they

are prepared by doing the preparatory information and activity sheets for those tests in *Rivers Chemistry*. Before going to the field, assign groups to conduct the specific chemical tests. Review relevant safety procedures.

7. Have students do the procedure for Student Activity 4.4: Determining the Benthic Macroinvertebrate Index for Your River or Stream. Indicate what technique you want students to use to pinpoint location (such as topographic map, compass, legal description, or river mile marker.) Indicate whether you want students to combine samples from different collections of habitats, or prepare an index for each sample. If you want students to keep samples separate, make sure you have sufficient collection buckets. If students mix them together, they should complete one set of indices; if they keep the samples separate, have them do a set of indices for each sample. If desired, have students make permanent collections of representative benthic macroinvertebrates.
8. As students carry out the activities in Lesson 4, observe and evaluate their individual performance on laboratory skills, recording your assessment on the Biological Field-Study Proficiency Checklist on page 209 of Appendix A, Assessment Tools.
9. Have students do Student Activity 4.5: Completing a Riverine Habitat Survey. If you want students to measure flow rate, use the specific student activity in *Rivers Chemistry* or *Rivers Earth Science*. If you want students to determine channel width, use the specific student activity in *Rivers Earth Science*.
10. If students will be doing the chemical tests for Student Activities 5.4 and 6.3, have them do them now.

Concluding the Lesson

1. In the classroom, have students complete Student Activity 4.4.
2. Have students do Student Activity 4.6: Determining the Diversity Index for Your River or Stream. For specific activities on logarithms, inverse signs, and other mathematical concepts, see *Rivers Mathematics*.
3. Have students compare their results and observations from Student Activity 4.5. Discuss and resolve any discrepancies. Using the habitat survey form in this activity, create an official class habitat survey by making a clean copy of the final tally, with all the responses agreed upon by you and the class. Make sure the form includes date, school, class, and teacher. You may also send copies to local environmental people who have helped your class. Place the completed habitat survey form in a labeled binder for future reference. Label the binder appropriately, so current and future students can retrieve information from it. Student Activity 4.5 includes an optional instruction that students create a children's nonfiction book about the local river or stream. For specific information on teaching nonfiction writing, see *Rivers Language Arts*.
4. If desired, have students, in groups or as individuals, make and display a permanent collection of aquatic macroinvertebrates.
5. Have students share and compare results and observations from all surveys and tests performed in conjunction with the field trip. Discuss and resolve any discrepancies. Use student responses as a basis for discussion of the field study. Prepare a final copy of the data for each index or test, and insert this into the labeled binder for future reference.
6. Ask students how their impressions changed as a result of their firsthand observations of the river or stream and its environment.
7. Have students discuss suggestions for any changes in procedure as a group, as a class, or with you individually.
8. If you plan to have a public or significant audience for the public meeting described in Student Activity 7.4, you may want to consider doing the first three steps in the Teacher Notes for Lesson 7 now. Then have students prepare appropriate publicity for the meeting.
9. Distribute and assign reading of Student Information 5.1: Oxygen in Water as homework.

Assessing the Lesson

1. Review and assess student performance on the skills demonstrated during this lesson, as recorded on Biological Field-Study Proficiency Checklist, in Appendix A, Assessment Tools.
2. Have students prepare a technical report on one of the areas of study covered during the field-site visit. (For specific teaching activities on technical writing, see *Rivers Language Arts*.)
3. Have students add to their river or stream collages.
4. Have students do Student Assessment 4.7: A River Diorama, in class or as homework. Here is a suggested scoring rubric:

Scoring Rubric

Score	Expectations
0	Diorama not completed. Paper not developed. Group does not work together.
1	Diorama completed. Animal and plant lists represent the habitat but show minimum effort. Paper completed, minimal but useful. Problems in sentence structure, spelling, and grammar. All members contribute something.
2	Diorama completed and shows much work. Animal and plant lists represent the habitat. Paper has simple detail and describes habitat, making reference to one index. Some problems in sentence structure, spelling, and grammar. All members contribute.
3	Diorama completed and shows much work. Animal and plant lists describe habitat well. Paper has detail and describes habitat well, referencing at least two indices. No problems in sentence structure, spelling, and grammar. All members contribute.
4	Meets all criteria for 3, plus paper elaborates on habitat and references at least three indices.

Extending the Lesson

1. If desired, have students (on the same or separate field trip) perform a flora and fauna survey, or a survey of floating phytoplankton and zooplankton. These activities take considerable work and special equipment, so they are not included in *Rivers Biology*. (If students will do several field-site surveys in one visit, point out that human activity scares away large animals, so students should survey them first. Likewise, activity in the water may disturb the benthic macroinvertebrates, though they would not move far. Plants, insects, amphibians, and most reptiles will not move, so they can be surveyed last.) These activities expand student survey and identification skills and enable students to do a more complete assessment of the riverine environment in Student Activity 7.3.
2. Visit a collection of local animals and plants. Local conservation agents and university biology departments may have or know of such collections.
3. Repeat the benthic macroinvertebrate survey or habitat assessment at a different river or stream site that has different characteristics. Compare results.
4. Have students study and observe periphyton. Explain that periphyton are any microscopic and macroscopic organisms that live attached to underwater surfaces. They may be attached to plants, decaying leaves, sticks, rocks, and other structures. Common periphyton include algae, sponges, hydra, planaria, and mollusks of various kinds. Students can use knives or spoons to scrape them from the surface of rocks, turtle shells, and fish scales.
5. Have students do a first-hand study of organism colonization of underwater structures. Have them place in the river or stream a Hester-Dendy

device. This device, used extensively to detect the zebra mussel, can be purchased from a biological supply house. Remove the device from the water periodically to observe the periphyton colonizing its surface. To observe the organisms, clip a microscope slide on the edges of the device so that the slide can also be colonized. To observe the organisms, remove the device from the water, remove the slide and place it in a clean jarful of river water. Attach a new slide, to the device, and return the device to the water. Return to the classroom to observe the sample under the microscope. Expect algae and protozoa in one to two weeks, hydra, sponges and other more complex organisms in a month. By collecting and observing a slide each month, students can observe patterns of how organisms colonize underwater structures.

Transparency Master

TEACHER TABLE 4-1
Benthic Macroinvertebrates in Sample
4 black fly larvae
1 crayfish
2 riffle beetle larvae
3 dragonfly larvae
8 sowbugs
15 mayfy larvae
1 water penny
2 fingernail clams
2 aquatic worms

Transparency Master

TEACHER TABLE 4-2

Model of Benthic Macroinvertebrate Pollution-Tolerance Index

Group 1 Most Intolerant		Group 2 Moderately Intolerant		Group 3 Fairly Tolerant		Group 4 Most Tolerant	
Stonefly larva	___	Caddisfly larva	___	Black fly larva	X	Aquatic worm	___
Alderfly larva	___	Mayfly larva	X	Midge larva	___	Leech	X
Dobsonfly larva	___	Riffle beetle larva	X	Sowbug	X	Left-hand/ pouch snail	___
Snipefly larva	___	Water penny adult	X	Right-hand/ other snail	___	Bloodworm midge larva	___
		Dragonfly nymph	X	Scud	___		
		Damselfly nymph	___				
		Cranefly larva	___				
		Crayfish	X				
		Clam/mussel	X				

Totals and Weighting

Sample 1

A: # of taxa =	0	# of taxa =	6	# of taxa =	2	# of taxa =	1
B: # of taxa × 1 =	0	# of taxa × 2 =	12	# of taxa × 3 =	6	# of taxa × 4 =	4

Sample 2

A: # of taxa =	0	# of taxa =	4	# of taxa =	2	# of taxa =	1
B: # of taxa × 1 =	0	# of taxa × 2 =	8	# of taxa × 3 =	6	# of taxa × 4 =	4

	Sample 1	Sample 2
C: Total Weighted Scores	22	18
D: Total Number of Different Taxa	9	7
E: Weighted Average # Taxa (total weighted scores/total of # of taxa)	2.4	2.6
Water-Pollution Tolerance Score	Good	Fair

Weighted Average Number of Taxa	Water Quality
1.0–2.0	Excellent
2.1–2.5	Good
2.6–3.5	Fair
Over 3.6	Poor

Transparency Master

TEACHER TABLE 4-3

Model of Diversity Index

Taxon (A)	Number of Organisms (B)	P (Orgs in Taxon / Total Orgs)	P (decimal form)	P Log 10	Log P (P Log 10 / .301)	Index Value (Log P·P)
Mayfly	15	15/38	.39	−0.4	−1.35	−0.53
Riffle beetle	2	2/38	.05	−1.3	−4.32	−0.22
Sowbug	8	8/38	.21	−0.68	−2.25	−0.47
Water penny	1	1/38	.03	−1.52	−5.06	−0.15
Crayfish	1	1/38	.03	−1.52	−5.06	−0.15
Dragonfly	3	3/38	.08	−1.1	−3.64	−0.29
Black fly	4	4/38	.10	−1.0	−3.32	−0.33
Fingernail clam	2	2/38	.05	−1.3	−4.32	−0.22
Aquatic Worm	2	2/38	.05	−1.3	−4.32	−0.22
Total	38					−2.58
Diversity Index (Inverse of Total of Index Values)						2.58

Diversity Index	Water-Quality Indications
< 1	Few taxa, some with many individuals; may indicate heavily polluted water
1–3	May indicate moderately polluted water
3+	May indicate relatively clean, unpolluted water

Name

STUDENT INFORMATION

4.1

Looking at the River

The field-based activities in this lesson form the heart of *Rivers Biology*. You will make observations and collect data that relate to living riverine organisms. Through this work, you can learn some of the organisms that inhabit the aquatic environment and the adjacent shoreline. You can also understand better how they indicate the diversity and the water quality of that river or stream ecosystem.

Biological surveys indicate both water and environmental quality, so their results may indicate whether chemical and other complex analyses are necessary. You may use the survey you conduct to characterize the present state of your local river or stream or to pinpoint problems. If your results show that your river or stream has good water quality, then your data can serve as a baseline for comparing data in the future. If results show impairment of the river or stream, then you, or others, may try to determine the cause and, ideally, work toward environmental and water-quality restoration. In such circumstances, the data provide values against which to measure improvement.

Making and Comparing Observations

You will be focusing on benthic macroinvertebrates for most of your site observations that indicate water quality. You can also garner information from looking at the riverine habitat directly. A comparison of the chemical and physical factors of a river with biological surveys of the resident organisms also provides much insight about the quality of life-sustaining factors of those waters. If you have access to data collected seasonally or annually, you can identify and quantify changes in water quality.

Some singular event may temporarily change the water quality and possibly cause the loss of less-tolerant organisms without having a significant effect on the main population of organisms. Such events may include increase or decrease in oxygen concentration, chemical spill, seasonal temperature change, and the addition of nutrients from a storm or flood. If those conditions exist on an ongoing basis, however, macroinvertebrates that cannot tolerate or outlast the condition will die, because they cannot move to a new location as fish can.

Organisms that can tolerate the adverse condition replace those affected by it, so the population and diversity in the ecosystem changes. A body of water reflects the long-range conditions being placed on it. For such reasons, the most useful biological analysis compares observations or testing results from one time or place with another set of data. Your teacher may be able to provide you with resources so you can do such comparisons with the data you collect.

Human Interventions Can Take a Toll

Human intervention has changed the biotic nature of almost every river; these changes have, in turn, altered the types and numbers of river-related plant and animal habitats. In your evaluation of a local aquatic habitat, you will be able to study how human factors have impacted that body of water. To do this, you might compare your results with those from a river that has had less human intervention.

Many types of human activity affect the flow and temperature of a river or stream which, in turn, affect aquatic organisms. **Flow** is the volume of water passing a particular site in a particular amount of time. Large organisms such as fish must have higher levels of flow than smaller fish or benthic macroinvertebrates. As a stream's flow slows, its temperature rises. Less oxygen can remain dissolved in warmer water than in colder water, so when low stream flow means higher water temperature, it also means lower oxygen levels in the water, sometimes too low for the survival of intolerant species.

If the banks of a stream or river are bordered with overhanging trees and brush, these plants shade the water, reducing direct heating from the sun. Human action may take away such natural vegetation, either by directly removing plants or by causing high variations in water flow that sweep away such vegetation and supporting soil.

Industrial and sewage uses of river or stream water may also raise water temperatures. For instance, power plants may use river water for cooling, then return it downstream to the river again but at a higher temperature. Because this warmer water has less oxygen, it can kill normal river organisms. Such problems, however, tend to affect relative few rivers, and relatively few stretches of water, compared with changes in bankside vegetation.

By far the most significant human factor in increased water temperature is alteration of stream flow. Humans remove water from the stream or river for irrigation, manufacturing, drinking water, and many other uses, particularly during dry times. This further exacerbates the effects of periods when water levels would naturally be low.

Many rivers and streams are controlled by dams that block the flow of river water, by deepening or straightening the stream channel, or by **levees** (LEH vees)—low ridges that keep the water within the river channel. A river or stream without such controls naturally has a diversity of types of habitats. The water may flow broadly but shallowly over stones, then form a narrow deep pool, then fall quickly over a drop to form a frothy mixture of air and water at the bottom, then curve around a bend where trees overhang the water on the inside curve. The structure of each habitat and the flow of water in that place interact to produce an environment that supports certain organisms. Each setting supports certain organisms, leading to a diversity of species along the river or stream. When dams, levees, and channels simplify and control the path of water, such diversity of habitat, and therefore, of species, tends to fall significantly.

Human alteration of stream flow has had a significant effect on the seasonal movement, or **migration,** of some species of fish, from oceans upstream to reproduce or spawn. The construction of dams impedes these natural migration paths. Even though utilities and public entities have constructed fish ladders to help fish bypass the dams and regularly transport large numbers of adults around the objects, salmon populations, for instance, have steadily decreased since the construction of large dams. These effects have also impacted other organisms that depend on those migrating species.

Though humans have designed many alterations of the river's path to reduce the likelihood of flooding of areas inhabited or used by people, in fact, some human activities actually increase the potential damage to nature that flooding can bring. For instance, in urban areas, construction and pavement—such as highways, buildings, sidewalks, and parking lots—have replaced wetlands and other land that once absorbed water from heavy rains and snow melt. Instead, that water now flows directly to the river, raising it to higher levels.

As these human actions increase the speed and quantity of water flowing to and into the river, they increase **erosion** (ih ROH zhuhn), the separation of soil and rocks that are subsequently carried into the river or stream. A swollen river or stream, moving with greater force because of its increase in volume, washes away more vegetation and soil from the land area adjoining the stream. Storm sewers, too, channel precipitation to the stream bed more quickly, accelerating the extreme nature of flooding. After the flood, water levels fall more than they naturally would, with no slower runoff from that water-soaked earth available to maintain water levels more supportive of aquatic life.

DRIFTWOOD

Over 1.7 million of the estimated 5 million to 30 million different life forms on Earth have been cataloged. Because hundreds of thousands of species may be extinct by the year 2000, the world has neither the scientists nor the time to identify the yet uncounted. Says University of Pennsylvania biologist Daniel Janzen, "It's as though the nations of the world decided to burn their libraries without bothering to see what is in them."

Pollution in the Water

Human activities may add materials to the water that impact its suitability for aquatic plants and animals. Such additions may include: sewage **effluent** (the watery materials resulting from sewage treatment); soil or sand from construction or channel alteration; agricultural runoff from pesticides, fertilizers, and animal manure; and outflow from fish growing. Human activity may also add solids or liquids to the water via dumping, accidents, or runoff.

Such materials mixing with water can suffocate animals and plants in the water. Any additions that affect odor, color, taste, clarity, or chemical composition of the water may have a negative effect on organisms. Some are visible, causing a scum, oil, or physical mess. Others are invisible, but may alter the chemistry of the water, encouraging some plants to proliferate rapidly, crowding out and reducing the oxygen and nutrients available to other species.

Often, public regulatory agencies manage the effect of such substances on the river by making sure they are significantly diluted. They allow into water only the amount of material that would not additionally affect the water or its organisms. **Point-source pollution** refers to pollution from a human activity detectable and controllable by a reasonable level of management. Point-source pollution comes from a pipe or other readily identifiable source. The primary type of point-source pollution is sewage discharge. Another type is material placed in the water by manufacturing. Agencies may trace a pollutant to a single source by testing water further and further upstream until the contaminant is not evident. For instance, they might trace high bacterial levels (another common form of point source pollution in rivers and streams) to a bad septic system in one or more houses or a cattle- or hog-raising operation.

Pollution whose human activity source cannot be readily identified is **nonpoint source pollution.** Common examples include: sediment moving off a farmer's field or a construction site; residue from a parking lot or highway; and acid rain from thousands of automobiles or businesses. If those materials enter a river or stream after a rain, they can greatly impact water quality and the lives of aquatic organisms, yet no single source can be blamed. Most point sources can be detected, but nonpoint sources are almost impossible to locate.

Human activity also sometimes inadvertently introduces destructive species to a riverine habitat. Most commonly, a large ship takes on water as ballast in one port, sails halfway around the world, then dumps the ballast water in another port. That ballast water may contain species native to one area. These species face no natural enemies in the new location and proceed to destroy native species directly, consume their food source, or take over their habitat, leading to the loss of native species. Other times, invasive shelled species attach themselves to a ship in port, then travel to another location and drop off to find new habitat for reproduction. Examples of nuisance aquatic life include the zebra mussel and Asiatic clam.

Preparing for Your Time on the River

Sample collection at a river or stream requires adequate advanced preparation and a selection of materials appropriate to the site chosen. For instance, biologists most often collect benthic macroinvertebrates in shallow streams with riffle bottoms. Larger rivers, silty bottoms, or extremely fast rapids require other sampling methods and may even affect results. So, familiarize yourself with the section of river or stream under study. The materials you will need depend on the characteristics of the river or stream, so someone will have to review maps of the river and visit the site. Your teacher will probably already have done that.

Always follow safety procedures for fieldwork, as described in Student Information 1.3. Never enter any stream without permission from the land owner (private or public). Never overlook precautions such as life jackets, first-aid kits, tow lines, and emergency transportation, even if they seem inconvenient. You must wear proper attire for fieldwork. Make sure you are aware of all potential dangers in the vicinity in which you will be working, such as poisonous plants and hazardous fauna. When selecting who will actually enter the stream to collect, you may consider the ability to swim as a determining factor. Review the safety contract before leaving for the field trip.

Questions

1. What might species diversity tell you about your particular river or stream community?
2. Describe how your local river or stream has been changed or controlled to meet human needs.
3. Identify two physical and two chemical human interventions that potentially could occur in your community's river or stream ecosystem. Discuss how you can recognize the effects of those changes.
4. How can familiarity with the site under study help you prepare better for doing a biological survey?

Name

Guide to Sampling Benthic Macroinvertebrates

When you collect biological samples from a river or stream, you should use techniques that provide the most accurate results with the least disturbance of habitat. Being on a river or stream is a very special experience for most of us. It is hard to believe at first, as you walk out in the water, that the homes of tiny, defenseless critters lay at your feet. If you kick through the water stirring up rocks and leaves, that action may disturb hundreds, sometimes thousands, of aquatic organisms. Sure, the mud washes away and the sand falls to the bottom, but, as the leaves drift away, so do the larvae of a myriad of insects and other organisms.

Though you must enter the stream in order to collect data, leave the stream as undisturbed as possible. The old environmental adage of walk across the forest but do not leave tracks holds equally well in a stream or river. Though others cannot easily see your tracks in the water, you will know you made them. Make gentle tracks in the water.

Why Do Sampling?

To help develop a picture of the health of a river or stream, aquatic scientists developed methodologies for sampling rivers for biological information. In *Rivers Biology,* you will use methods scientists actually use in their work. The method described in *Rivers Biology* for sampling and quantifying organisms in a stream or river is based on work developed by the Izaak Walton League of America's Save Our Streams (SOS) program, then adapted by the Illinois Department of Natural Resources.

States may have different species of benthic macroinvertebrates. Though sampling for benthic macroinvertebrates is basically the same in any setting, many states have developed a procedure that more specifically reflects their conditions. Your teacher may have checked with your state's department of conservation or natural resources and adapted the material in *Rivers Biology* to correspond to a more local protocol.

General Sampling Considerations

Like all scientific methodologies, the accuracy of sampling depends on following a procedure precisely without making mistakes, so others can produce constant data as a check of that precision. For most accurate and responsible results, always carry out biological sampling with the least disruption to the communities being sampled. To do so, heed the following guidelines.

1. Sample at even flow or, at least, stable water discharge conditions. The stream should be as typical as possible, so do not sample just after a water release or other human activity. Such activity agitates the water, dislodging or covering organisms.
2. For long-term investigations, sample at relatively the same time (or times) each year. When the water level is low and water temperature is relatively high (typically in late summer to early fall), organisms have less oxygen, so their numbers may differ significantly from periods of higher water and lower temperatures.
3. When doing more than one survey on a field trip, plan your sampling sequence carefully. During *Rivers Biology,* you will probably survey only benthic macroinvertebrates. If your teacher indicates that you will sample other organisms also, sample fish first, then macroinvertebrates, with habitat study last. Fish move away from activity in the water quickly; such activity will also cause the macroinvertebrates to hide, though more slowly. On land, plants and most animals, except the big ones, are little affected by activity in the stream, so you can sample them last.
4. When entering the stream or river, start downstream, moving upstream to collect.
5. Make sure you and your classmates do not all walk through the same area.
6. The type of environment being sampled dictates the types of equipment and techniques that will yield the most accurate results. Determine whether the bottom of the stream bed where you will be sampling tends to be rocky or muddy. Follow the procedures that most closely correspond to your field site.
7. Developing good technique takes practice. If your teacher suggests, spend time practicing use of the sampling nets before doing actual sampling in the field.

How to Sample a Rocky-Bottom Stream

Use this method of sampling in areas that have faster flow, disturbed by flowing over rocks, usually shallow. Many species of benthic macroinvertebrates thrive in such habitats.

1. Choose an area to test that is at least 30 meters (100 feet) from any human modification of the river or stream. Look at the riffles, areas of disturbed water with faster flow over fist-size rocks. Select a representative riffle, with swift flows and rocks that are rough, flat, or large enough to serve as a habitat for macroinvertebrates. In this riffle area, select a sampling area 1 meter by 1 meter square. Because sampling in deep water is difficult, select an area no more than 60 cm (2 feet) deep. Remember, approach this area from downstream, moving upstream.

2. To collect benthic macroinvertebrates in these environments, you will need a kick net. A **kick net,** shown in Figure 4-1, is a seinelike net used to capture benthic macroinvertebrates and other small animals. A kick net is so named because someone goes above the net and kicks loose the rocks and leaf litter to dislodge any organisms, which then are captured in the net. Place the kick net at the downstream edge of your selected sampling area. Have two students hold the net in the water at about a 45-degree angle with the surface. The bottom of the net must fit tightly against the bottom of the stream, so no organism can escape. Do not allow water to flow over the top of the net. The net holders must be aware of the position of the net bottom at all times. If necessary, place a rock or two on the edge of the net to keep the net on the bottom.

DRIFTWOOD

"All too often, we are too busy to just stop and look around. Someday, we will be sorry we didn't take one moment to cherish the gifts of nature."

Unknown

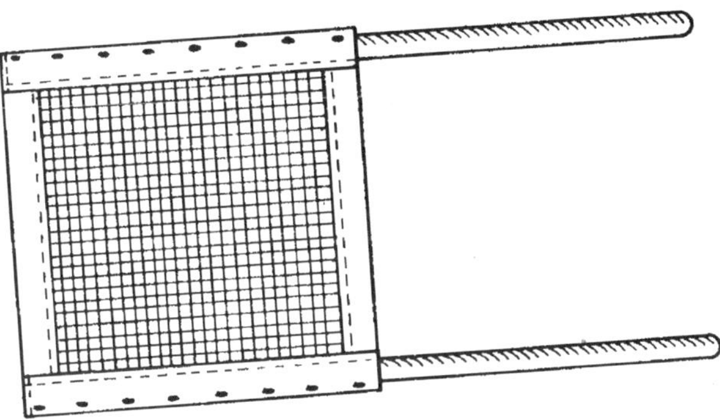

Figure 4-1: Kick Net

3. The goal of this activity is to dislodge all the small macroorganisms from the stream bed and rocks in your sample area. To do so, be systematic and heavy-footed. Dislodge organisms from all surfaces within your sample area. To do this, have a third student brush his or her hands over all these surfaces. Turn over, separately, each moveable rock, stick, or piece of trash. As you handle each item, rub it with your hands. Then rub your hands over the hole left by the item to disturb animals hiding there.
4. Hand each item to a fourth collector, who should hold it in the water bucket and brush it to remove clinging organisms. Return each item (except trash) to its original position.
5. Stir the stream bed with feet or a stick over the entire one-cubic-meter area. Continue this disturbance for at least 60 seconds.
6. Remove the kick net from the water by moving it upstream in a scooping motion, so no organism trapped on its surface is lost. The force of the water will hold the captured organisms on the net surface. Remove the big organisms and the twigs and rocks caught in the net, checking for small, hard-to-see macroinvertebrates. Record any amphibians, fish, or reptiles caught in the net; then release them.

7. To remove the benthic macroorganisms, roll the kick net into a cylinder and place one end in the collecting bucket filled with stream water. Pour or spray stream water over the net to wash the organisms down into the bucket. Avoid overfilling the container; this would wash organisms out.
8. Pour off the water, and place the contents of the bucket in a white collecting pan or on a white sheet for sorting and identification.
9. If the net comes up empty, repeat with a different net angle, and make sure it stays against the bottom.

Sampling a Muddy-Bottom Stream

Use this method of sampling in stream or river sites where the stream bottom is covered with mud, silt, or sand. Muddy-bottom waters feature three different environments for macroinvertebrates, each supporting different types of organisms. Accurate sampling, therefore, requires collecting from each of these environments. These habitats are:

- snags, submerged logs, tree roots, aquatic vegetation, and noticeable human debris
- gravel, leaf piles, and other organic debris (such as twigs and sticks)
- sandy and muddy bottom substrate.

Usually, the most pollution-sensitive organisms will be found in habitats of the first type, such as submerged logs. The most pollution-tolerant live in sandy-and muddy-bottom substrates.

The goal of this activity is to dislodge all the small macroorganisms from the stream bed and from submerged and floating objects in your sample area. First, sample areas in which the most pollution-sensitive organisms live, such as snags and tree roots. Then sample the second area, such as gravel or leaf litter. Sample the bottom substrate last.

1. Choose an area to test that is at least 30 meters (100 feet) from any human modification of the river or stream. The area you select should have all three collecting locations in the same general area, the closer to each other the better. Sampling multiple locations within the area will assure that every possible macroinvertebrate taxon has been located. Some collecting protocols define how large an area to test; this one does not. The collected area should be as diverse as possible.
2. First, sample snags, logs, and vegetation. Walk along the stream for several meters scraping the surface of snags, tree roots, logs, and vegetation with a net to collect benthic macroinvertebrates. Collecting nets used for gathering organisms in water called **dip nets,** come in two kinds, as shown in Figure 4-2. A **D-net** is long-handled, with netting in D shape. A **triangular net** has netting in a triangle shape. These nets are flattened on one side, so users can collect organisms in the water by skimming the bottom of the stream.

Figure 4-2: D-net and Triangular Net

3. Drag the collecting net through submerged vegetation and piles of debris. Pick up debris and place it in the net. If you can enter the water safely, pull loose any snags or drag the net along their surface.
4. With the net placed downstream to capture drifting animals, use any safe activity to dislodge organisms by moving the logs, snags, or vegetation from their location. Have another student hold each item over a collecting bucket. Have a third student rub his or her hands over the surfaces of each item to dislodge any organisms into the bucket. Return each log or other item (except trash) to its original position in the stream.
5. Next, sample the gravel, leaf piles, and other organic debris. Walking in or along the stream, use the net to scoop up surface gravel, leaves, and other debris that is laying on the bottom of the stream. The amount collected will depend on the composition of the bottom and how heavy the net becomes.
6. Spread out the entire contents of the net on a white surface, or place them in a collecting bucket filled with stream or river water.
7. Carefully sort out the debris, looking at each piece for benthic macroinvertebrates, Examine each piece of debris carefully; some organisms blend extremely well into their habitat materials. Moreover, when exposed to sunlight or wind, water organisms move toward places with water. With forceps, eyedropper, spoon, or small paintbrush, transfer the captured invertebrates into a collecting pan.
8. Place the gravel, leaves, and other organic debris back into the water.
9. Finally, sample the sandy or muddy substrate. Place the net on the bottom of the stream or river. Drag it hard across the bottom. You may also kick the mud above the mouth of the net, allowing the current to carry

dislodged animals into the net. Sweeping the net through a disturbed area will capture animals that are being carried away from your activity. If large amounts of mud have accumulated in the net, rinse most of it out by moving the net back and forth through the water.

10. Spread out the entire contents of the net on a white surface, or place them in a collecting bucket filled with stream or river water.

Questions

1. What do you know about the stream bed conditions at the site where you will collect benthic macroinvertebrates? What collecting procedures do you think you will use? Support your answers.
2. When you go to your local river or stream to collect, you will want to use a least-impact procedure. In your own words, make a simple set of steps for collecting at a stream or river while making the least impact on organisms in the water.
3. Why do you want to approach the collecting site from downstream?
4. Draw and label all the collecting equipment you plan to use for sampling your local river or stream.

Name

Riverine Habitats

As part of your study of the biological health of your river or stream, you will perform a habitat assessment. A **habitat assessment** is a survey of a particular site, ranking its suitability as an environment for plants and animals. Such a survey examines specific aspects of the stream environment, referred to as **parameters** (puh RA muh ters). Many of the parameters are physical, but others may include human factors and aesthetics.

River Habitat Assessment

Physical habitats are critical in maintaining the integrity of an aquatic ecosystem. Physical characteristics of a river or stream environment enhance or limit the life forms found in that water. For instance, mud-bottomed streams generally support fewer life forms than rocky streams. River or stream sites where water drops quickly, forming rapids and mixing with oxygen, better support aquatic life forms dependent on significant oxygen levels.

Biologists and other researchers identify physical characteristics of a specific stream or river in order to understand how these conditions influence the suitability of this habitat for various organisms. Sometimes they discover limitations or degradation caused by human activity; then they can recommend action to remediate or improve those conditions.

A riverine habitat survey looks at specific physical components of the stream environment, specifically the channels, bottom, stream banks, and watershed. Parameters generally covered are temperature, turbidity, siltation, velocity, depth, cover, pool and riffle sizes, riparian vegetation, bank stability, and land use. Some surveys include chemical and biological evaluations. The survey you will perform includes the chemical and biological parameters you can observe easily without special equipment or calculations.

To maximize usefulness of data from a habitat assessment, biological survey, or other water-related test, the data report should identify the precise location of the field site. Investigators use topographic maps, compass, and legal descriptions to pinpoint the location exactly. The investigator should also note weather conditions for the current day, and for preceding days when such information is available. Such variables can influence test results significantly. For instance, if a big rain has recently occurred, the water level will probably be higher, as will its oxygen content, but so might be levels of some contaminants, including sediments.

River or Stream Characteristics

Stream depth and flow may be the most important habitat parameters. They dictate the volume of water and habitat available to provide life support for organisms, such as shelter, food, and oxygen. Stream flow of greater than 0.15 m^3 [5 cubic feet per second (cfs)] in warm-water streams and rivers, or 0.06 m^3 (2 cfs) in cold-water streams, provides excellent water for supporting a stable aquatic system.

To estimate average stream depth, researchers measure the maximum depth of both riffles and deep portions of the stream. A stream with an average depth of greater than 60 cm (24 in.), with riffle depth adequate to allow free passage for fish and shelter for feeding, is excellent. Cold-water streams meeting this standard provide excellent habitat for trout and other cold-water fishes. Streams less than 30 cm (12 in.) deep may provide excellent habitat for invertebrates but may not be adequate for fish.

Stream Channel Characteristics

The ways in which water flows in a river or stream influences its suitability as aquatic habitat. Water may flow as riffles, pools, runs, or bends. As described earlier in *Rivers Biology,* a riffle is an area of disturbed water with faster flow, cobble, and usually, large invertebrate populations. A **pool** is an area of slow, quiet water that is spread out. A **run** refers to an area of moving but deeper water. Even though a deep run may have large rocks below, the surface water is quiet.

Most research supports the assumption that a stream with a mixture of riffles, pools, and bends contains better habitats for aquatic development than a straight or uniform stream. So, as an indicator of habitat health, some surveys include calculation of a ratio of pools to riffles, or, in gently-sloped streams that may not have riffles, a ratio of runs to bend. In such slower streams, bends with undercut banks provide important habitats conducive to aquatic life.

The **channel** of the river or stream is the portion of the stream bed in which the deepest, fastest, portion of the river or stream flows. Channel width, depth, **gradient** (steepness), and roughness determine the volume of water that can pass down the river or stream at a given time, known as **channel capacity.** Over time, channel capacity adjusts to the size of the watershed and changes in watershed vegetation. When channel capacity is exceeded, unstable areas are likely to erode, causing an increased deposit of soil in the water. When investigating whether channel capacity has been exceeded, researchers look for deposits of soil in the stream bed and for plant debris hung up on bank vegetation. Extreme stream velocity during high water can also impair habitat by scouring away soil.

Researchers consider soil deposits in the stream channel as indicators of erosion in the watershed and in the river or stream. Deposits usually occur on the downstream side of rocks and other objects that deflect water flow. Deposits may also take place on the inside of bends, below channel constrictions, and where the stream flattens out. On lower, flat stretches, deposition occurs during peak flows. The growth and appearance of new sand or silt bars indicates upstream erosion.

Substrate Characteristics

Materials at the bottom of the stream, on the bottom substrate, provide habitat to support aquatic organisms. Different organisms are adapted to living in different substrates, so the most diverse rivers and streams have a bottom substrate that includes a variety of substrate materials and habitats.

In determining the character of a substrate, researchers categorize its particles by size. Silt, clay, and mud are the smallest. **Gravel** refers to rock from three millimeters to five centimeters in diameter. Rocks from five to 30 centimeters are called **cobble,** with anything bigger being called a **boulder.** If a substrate is only solid rock, it is exposed **bedrock,** the solid expanse of material that forms a layer of rock below the surface materials.

In a habitat assessment, researchers assign percentages to each type of substrate present. The higher the percentage of cobble or small rocks, the greater amount of habitat available for benthic macroinvertebrates. A habitat with 60 to 90 percent cobble contains many places for small organisms to hide and live. In addition to rock substrate, downed trees, clumps of leaves, exposed tree roots, overhanging banks, and aquatic vegetation provide a temporary substrate that forms important habitats for aquatic life forms. Sand, small rocks, and mud bottoms, on the other hand, provide few habitats; so, fewer animals are adapted to living in such settings.

When examining the substrate, researchers also look at the embeddedness of its particles. **Embeddedness** refers to the degree to which the substrate particles are covered by the surrounding soil or sediment. Embeddedness is a partial measurement of how much soil has moved into the stream or river, a process called **siltation.**

Stream Bank and Watershed Conditions

The stream bank is the land adjoining a stream, composed of the upper bank, lower bank, and bottom; it includes from the break in the general slope of the surrounding land on one side of the stream to the low-water line. The **upper bank** is the land area from the break in the general slope of the surrounding land to the normal high-water line. This area is usually **vegetated** (covered with plants); at extreme high water or flood times, it is covered with water. The **lower bank** is the area from normal high water to the low-water line. The lower bank has few plants and often is underwater after a rain. The constant coverage by water keeps many plants from becoming established.

Vegetated streams that slope to the water relatively gently are less likely to erode than steep, unvegetated stream banks. The extensive root systems of riverside plants hold soils and keep the banks from caving in or being washed into the water. Researchers usually determine the characteristic and suitability of stream banks for supporting aquatic habitats by measuring the percentage of the bank vegetated. If soil detaches and moves into the stream, **bank erosion** takes place. Bank erosion impacts aquatic organisms downstream, because it allows sediment to enter the water. That sediment can choke aquatic life, add nutrients, and cover habitat, thus lowering the water quality. Well-vegetated banks are a significant component of a high-quality aquatic habitat.

Watershed is the total land area from which water flows toward a common stream or river. The shape, mineral components, and use of that land may influence the suitability of the stream or river as aquatic habitat. The steepness, composition, and amount of vegetation on watershed land determines its erosion potential. **Erosion potential** refers to the ability of soil to detach or move into the stream. Erosion potential includes both natural and human-made factors that affect the likelihood of siltation.

Soil that erodes, moving into the stream, increases the amount of solids in the water, which may also increase the lack of clarity, or **turbidity,** of the water. Turbidity reduces the amount of sunlight entering the water, which, in turn, reduces plant growth and raises water temperatures. Highly turbid water supports certain species of fish and macroinvertebrates; clean, clear water supports others.

Siltation also affects acidity and, thus, the water's ability to support aquatic organisms. If sediment from limestone rock, for instance, enters a river or stream, the water's acidity will drop, which may endanger some organisms. Where mining operations have brought underground materials to the surface, water runoff over those materials may transform the stream or river into water so acidic that few, if any, organisms can survive.

Cultural Factors

Human actions that impact rivers and streams are considered **cultural factors.** Humans have affected nearly all rivers and streams in some way. Such changes often impact the suitability of the stream or river as an aquatic habitat. The factors fall into several categories, of which the two most significant are probably land use and **hydrologic modification** (change in the course of water at some point in the hydrologic cycle).

Land-use factors that impact the riverine habitat include residential, agricultural, and industrial uses. In the habitat assessment, you will note any such conditions, as well as current construction, roads, parking lots, or other paved areas. As an observer at the water's edge, you may not always notice the impact of such factors on your river or stream. Recognition may require repeated collection of river and stream data over time. For example, removal of native vegetation, destruction of wetlands, and paving of streets and parking lots increases runoff and decreases groundwater recharge. The only way to notice such changes, however, might be via systematic observations that reveal flow increases or significant variation in flow patterns.

Other land-use factors that influence riverine habitats include point sources of water pollution, such as wastewater discharge from industrial and agricultural operations and from septic systems directly into the river or stream. Along your river or stream, you may be able to identify possible point sources of pollution. For instance, you might find drainage pipe along the bank, a leaking septic tank in a house backyard, or drainage from a cattle or pig lot.

Once a substance enters the water, it may influence all or part of the habitat downstream. Several of the more common sources of such substances include feedlots, septic systems, mine seepage, parking lots, new construction, golf courses, areas of high fertilizer use, and city streets. Sources of nonpoint source pollution are difficult to determine, particularly without doing upstream field site visits. So you just identify materials and conditions you observe. The most common indicators of nonpoint pollution are trash, oil slicks, odor, stains, and low pH. Proper management techniques can control almost any source of pollution. Such factors are the most controllable cultural factors.

Habitat assessments also note any observable hydrological modifications of the river or stream. Dams provide massive control of the path and power of the river. Humans construct dams to help control navigation, provide water and electricity, build recreation reservoirs, and help with flood control. Many dams were built before concerns over habitat surfaced; their creation did, however, affect aquatic habitats. They created long runs by straightening rivers and streams, drained wetlands, and served as barriers to fish migration. They increased erosion in some areas, altered water temperature, and reduced flow in other areas. Other human modification include channel-building, levees, and water removal for human use (via pipeline or ditch).

The results of habitat assessments may provide information useful in making appropriate future industrial, agricultural, commercial, and residential planning decisions; less often, they may even spur changes in current land-use and hydrology activities that reduce riverine habitats.

Aesthetic Characteristics

Each river or stream has aesthetic value not necessarily related to its ability to support aquatic life. People's perceptions of what constitutes a desirable body of water—what gives it **aesthetic value**—is important to how they protect and nurture the river or stream. Aesthetics includes not only the river or stream's visual appearance but how it smells, feels, and even sounds to the observer. Taste can come into play if the community uses the water for drinking.

A habitat assessment asks the observer to rank the smell of the river or stream and indicate the presence of any nonriverlike smells. Possible inferences of such smells, as suggested by the Illinois RiverWatch Network, are provided in the nearby table.

TABLE 4-1

Water Odors

Odor	Possible Inference if Odor Present
None	Good water quality
Sewage	Human waste (if so, do not enter)
Chlorine	Overchlorination by water treatment plant
Fish	Presence of excessive algal growth or dead fish
Rotten eggs, sulfurous	Sewage pollution, underground thermal pools
Petroleum	Oil spill from storm sewer or other

Evaluation of visual appearance includes the water itself and the surrounding land. Abnormal water appearance usually indicates some human-made interference. Clearer water is generally associated with cleaner rivers, although some rivers are naturally turbid, and water that looks clear may still have high levels of toxins or infectious organisms. Typical inferences are presented in Table 4-2.

TABLE 4-2

Possible Significance of Water's Visual Appearance

Visual Appearance	Possible Inference of Visual Appearance
Clear	Good water quality
Milky	Industrial pollution or alkaline conditions
Foamy	Phosphates; agricultural fertilizer runoff
Light brown, turbid	Erosion
Dark brown	Excess decaying organic material
Reddish	Excess iron
Oily sheen	Petroleum products or natural oils
Greenish	Algal blooms from excess nutrients

The single most important aesthetic value placed on a river or stream is probably naturalness, its apparent status as untouched by human influence. People want to look out over the water and see wild things, lonely spots. They want to hear only natural sounds. With much of the country urbanized and traversed by motor-driven vehicles, such **naturalness** becomes more precious each day.

Visual Biological Survey

Field biologists perform many different types of surveys of the plants and animals in the habitats they study. They compare survey results with other habitat observations and test results to learn whether the observations reflect long-term or short-term conditions. (Long-term or extreme conditions are more likely to have affected the organism populations.) So a habitat survey process often involves at least simple visual surveys of the amphibians, waterfowl, reptiles, mammals, fish, aquatic plants, and algae at the site.

Macroinvertebrate Survey

Benthic macroinvertebrates, as you know, provide a particularly strong indication of the health of an aquatic environment. So, habitat surveys generally include observation of such organisms, for comparison with other observations. *Rivers Biology* includes performing a more detailed macroinvertebrate survey at your site, but this one provides a quick set of observations that may spark insights into the habitat.

Questions

1. If you could just look at five or six habitat variables, which would you use to relate the true condition of your river or stream?
2. Discuss plants as an indicator of water quality.
3. Give your aesthetic impression of your local river or stream, if you have visited it. If not, apply this question to a body of fresh water you have visited. Use more than visual descriptions.
4. List cultural factors that you hypothesize may influence your local river or stream site.
5. List nonpoint pollution sources you think affect your local river or stream site.

Name

STUDENT ACTIVITY 4.4

Determining the Benthic Macroinvertebrate Index for Your River or Stream

Purpose

To determine the benthic macroinvertebrate water tolerance index for a field site at your local river or stream and to use it as a measure of habitat health.

Background

Now that you understand the value of the benthic macroinvertebrate pollution-tolerance index, your teacher may have you practice it with a sample set of data before you collect actual benthic macroinvertebrates at your field site and produce an index of those organisms as an indicator of water quality. Do any habitat survey after the benthic survey.

Your stream or river site may have predominantly one type of substrate, either a rocky bottom or a muddy or sandy bottom. Use the collection technique described here for the substrate type that most closely corresponds with

Materials

Per student

- life jacket and tow line (for deeper-water sites)
- rubber waders or boots (for shallower-water sites)
- protective gloves (if water may have contaminants)
- clipboard
- journal
- Student Information 4.2: Sampling Benthic Macroinvertebrates

Per group

- kick net (for rocky-bottom stream or river)
- D-net or triangular collecting net (for muddy-bottom stream or river)
- shallow white pan
- white sheet, oilcloth, or tarp
- 2–3 collecting buckets
- rinsing bucket
- forceps
- eyedropper
- spoon
- 2-cm (1-in.) paintbrush
- small watercolor brushes
- laminated Benthic Macroinvertebrate Pollution-Tolerance and Field Identification Reference Sheet (Figure 3-1 from Student Information 3.3)
- macroinvertebrate identification guide
- hand lens or dissecting microscope

Per class

- topographic map or other material for determining site location
- first-aid kit
- bottled water for washing
- soap and towels
- large plastic bags for collecting waste

Optional

- variety of collecting bottles and clean jars with lids (for permanent collections)
- alcohol (70% ethyl) for preservation

For the teacher

- overhead transparency of Teacher Table 4-1, Benthic Macroinvertebrates in Sample
- overhead transparency of Teacher Table 4-2, Model of Benthic Macroinvertebrate Pollution-Tolerance Index

your site. If your site contains both types of substrates, try to sample them both.

Benthic macroinvertebrates generally prefer to live in the shelter provided by rocks. In muddy or sandy bottoms, which have few or no rocks, benthics may take shelter in other habitats. These may include leaves, twigs, bank overhangs, suspended roots, logs debris, or trash. So, try to sample several different sites in your collecting area. This will increase the diversity index, because your chances of finding a variety of individuals rise.

Procedure

1. In your journal, record time, date, and location. (Use the location identification technique your teacher indicates.) Record current weather and water conditions, and note weather conditions for previous days, to the extent you are aware of them.
2. Put on safety equipment.
3. Look at the bottom of the stream or river at your field site. If it has a rocky bottom, follow Part A. If is has a muddy bottom, follow Part B. If the stream contains both a riffle and a muddy bottom, do both parts. You can combine or keep separate the samples collected from both types of sites. The overall results of merging such samples would provide a representative sample of the overall stream, not of a specific stream habitat.

PART A. Sampling a Rocky-Bottom River or Stream

4. Choose an area to test that is at least 30 meters (100 feet) from any human modification of the river or stream. Select a representative riffle, with swift flows and rocks rough, flat, or large enough to serve as a habitat for macroinvertebrates.
5. From within this riffle, select a sampling area 1 meter by 1 meter square. You do not have to mark it, but make sure to examine all the rocks and bottom within this approximate area.
6. Wash the kick net to remove any organisms from previous collecting. Rinse and fill your collecting and rinsing buckets with clean river water.
7. Place the kick net at the downstream edge of your sampling area. Have two students hold the net in the water at about a 45-degree angle with the surface. The bottom of the net must fit tightly against the bottom of the stream, so no organism can escape. Do not allow water to flow over the top of the net. If necessary, place a rock or two on the edge of the net to keep the net on the bottom.
8. Dislodge organisms from all surfaces within your sample area. To do this, have a third student brush his or her hands over all these surfaces. Turn over separately each moveable rock, stick, or piece of trash. Rub your hands over the rock or other item. Then rub your hands over the hole left by the rock to disturb animals hiding there.
9. Hand each item to a fourth collector, who should hold that item in the water bucket and brush it to remove clinging organisms. Return each item (except trash) to its original position.
10. Stir the stream bed with feet or a stick over the entire 1-square-meter area. Continue this disturbance for at least 60 seconds.
11. Remove the kick net from the water by moving it upstream in a scooping motion, so no organism trapped on its surface is lost. The force of the water will hold captured organisms on the net surface. Remove big organisms, twigs, and rocks caught in the net, checking for small, hard-to-see macroinvertebrates. Record any amphibians, fish, or reptiles caught in the net; then release them.
12. To remove the benthic macroorganisms, roll the kick net into a cylinder and place one end in the collecting bucket. Pour or spray water over the net to wash the organisms down into the bucket. Do not overfill the container.
13. If the net comes up empty, repeat with a different net angle, and make sure the lower edge of the net is held against the riffle bottom.
14. Repeat Steps 5 through 13 in a close but different riffle, with

a somewhat slower current, using a separate collection bucket. You may even repeat these steps for a third riffle. For the indices in *Rivers Biology,* your teacher may have you combine the samples, or consider each collection separately. If you mix them together, complete one pollution tolerance and one diversity index. If you keep the samples separate, complete these indices for each sample.

PART B. Sampling a Muddy-Bottom River or Stream

15. Choose an area to test that is at least 30 meters (100 feet) from any human modification of the river or stream. Select a representative flowing shallow area.
16. From within this area, select a sampling area 1 meter by 1 meter square. You do not have to mark it, but make sure to examine all the rocks and bottom within this approximate area.
17. Wash the collecting net to remove any organisms from previous times. Rinse and fill your collecting and rinsing buckets with clean river or stream water.
18. Walk across or upstream along the stream for several meters in your collecting area looking for places organisms might select, scraping the surface of snags, tree roots, logs, and vegetation with the collecting net. Have the net facing into the flow of the water at all times to keep the organisms in the net. In order to avoid losing animals, empty the net into a collection bucket frequently.
19. Drag the net through any submerged vegetation and piles of debris. If you can enter the water safely, pull loose any snags, or drag the net along their surface.
20. With the collecting net placed downstream to capture drifting animals, use any safe activity to dislodge the organisms. If you can, have one student remove the logs, snags, or vegetation from their locations. Have another student hold each item over a bucket. Have a third student rub his or her hands over the surfaces of each item to dislodge any organisms into the bucket. Return each log or other item to its original position. Empty the net into a collection bucket frequently.
21. To sample the gravel, leaf piles, and other organic debris, walk in or along the stream using the collecting net to scoop up surface gravel, leaves and other debris. Try to sample as many different sites in your collecting area as possible. Empty the net into a collection bucket frequently.
22. To sample the sandy or muddy bottom, place the collecting net on the bottom of the stream or river. Drag it hard across over a square meter of the bottom. Rinse off extra mud by moving the net back and forth through the water.
23. Another pair of students should kick the mud above the mouth of the net, allowing the current to carry dislodged animals into the kick net. Rinse off extra mud by moving the net back and forth through the water. Using all the techniques will give a better sample and a better view of the stream's diversity because you have looked at all the many places animals might live.
24. To test the extent to which the sample taken at this area is representative of the stream, repeat Steps 15 through 23 for another site a hundred yards upstream, using a separate collection bucket. You may even repeat these steps for a third site. For the indices in *Rivers Biology,* your teacher may have you combine the samples, or consider each collection separately. If you mix them together, complete one pollution tolerance and one diversity index. If you keep the samples separate, complete these indices for each sample.

PART C. Identifying the Samples

25. When you have completed your collection, dump the contents of the first collecting bucket into a shallow, white pan. If you have a large volume of materials, use a larger white surface, such as an oilcloth, sheet, or tarp. Use forceps, brushes, eyedroppers, spoons, and other devices to

pick through the leaves, twigs, and other debris to remove all the benthic macroorganisms into containers filled with river water. Look carefully because many are quite small or fast-moving. Riffle beetles and snails become immobile when disturbed, resembling pieces of debris or rock.

26. Identify the organisms, using appropriate keys, hand lens, or dissecting microscope. Identify the macroinvertebrates by order, record the numbers, and the taxa. The word *taxon* is used here to refer to an apparent species. A mayfly is one taxon. If you have two different species of mayflies, you have two taxa. If two organisms look as if they are the same species, then you have one taxon. Count how many macroinvertebrates are in each taxon collected. Place each total on your data sheet.
27. If you are making a permanent collection of benthic macroinvertebrates, follow your teacher's instructions for placing one of each species into a small container in alcohol.
28. Return to the river or stream any organisms you have not added to your permanent collection, as close to their original locations as reasonably practical.
29. Repeat Steps 25 through 28 for each sample you have collected.
30. Wash the nets and containers in the river or stream to remove any mud or leaf litter.

PART D. At the Classroom

31. When you return to the classroom, clean or wash each item used. Wash the nets and place them in a position to dry. Store the nets only when completely dry.

Observations

River or Stream ________________ Location ______________________________

Date __________________________ Time _________________________________

Weather Conditions ____________ Weather Previous 24 hrs _______________

Water Conditions ______________ Air Temp ____ °C Flow Rate ____ m/sec

School ________________________ Investigators __________________________

DATA TABLE 1: Survey of Benthic Macroinvertebrates

Sample # _____

Group 1 Most Intolerant (Best Water Quality)	**Group 2 Moderately Intolerant**	**Group 3 Fairly Tolerant**	**Group 4 Most Tolerant (Worst Water Quality)**
Stonefly larva ___	Caddisfly larva ___	Black fly larva ___	Aquatic worm ___
Alderfly larva ___	Mayfly larva ___	Midge larva ___	Leech ___
Dobsonfly larva ___	Riffle beetle larva ___	Sowbug ___	Left-hand/pouch snail ___
Snipefly larva ___	Water penny adult ___	Right-hand/other snail ___	Bloodworm midge larva ___
	Dragonfly nymph ___	Scud ___	
	Damselfly nymph ___		
	Cranefly larva ___		
	Crayfish ___		
	Clam/mussel ___		

Totals & Weighting

A: # of taxa ___	# of taxa ___	# of taxa ___	# of taxa ___
B: # of taxa × 1 = ___	# of taxa × 2 = ___	# of taxa × 3 = ___	# of taxa × 4 = ___

DATA TABLE 2: Benthic Macroinvertebrate Pollution-Tolerance Index

	Sample 1	**Sample 2**	**Sample 3**	**Sample 4**
B: Weighted Scores				
Group-1 # of taxa × 1				
Group-2 # of taxa × 2				
Group-3 # of taxa × 3				
Group-4 # of taxa × 4				
C: Total Weighted Scores				
D: Total Number of Different Taxa				
E: Weighted Average # Taxa (total weighted scores/total # of taxa)				

Final Index Average Total Value
(total weighted average # taxa/# samples) = ____________

Water-Pollution Tolerance Score = ____________

Calculations

32. Transfer the data for your Sample #1 from your journal onto your own Data Table 1, Survey of Benthic Macroinvertebrates.
33. Total the number of taxa in each tolerance group. Enter the result in Data Table 1, Line A.
34. For each tolerance group, multiply the taxa total by the corresponding weighting factor (the group number). Enter the resulting weighted score in Data Table 1, Line B.
35. On your own Data Table 2, Benthic Macroinvertebrate Pollution-Tolerance Index, add the weighted scores for all groups in this sample. Enter the result on Data Table 2, Line C.
36. Add the unweighted number of taxa for all groups in this sample. Enter your result on Data Table 2, Line D.
37. Divide the total weighted score obtained in Step 6 by the total number of taxa obtained in Step 7 to get the weighted average number of taxa. Enter the result on Data Table 2, Line E.
38. If you or the class have taken multiple samples from the same river or stream, repeat Steps 32 through 37 for each sample, creating a new table like Data Table 1 for each.
39. Average the results for all samples to develop a final index average value.

Analyses and Conclusions

Weighted Average Number of Taxa	Water Quality
1.0–2.0	Excellent
2.1–2.5	Good
2.6–3.5	Fair
Over 3.6	Poor

1. Using the final index average value for this site (or the weighted average number of taxa if using only one sample) and the preceding chart, rate the water at this site from excellent to very bad.
2. From your results, what can you conclude about the quality of the water at this site?
3. If you collected at the riffle and at a mud bottom, where was the greatest number and types of organisms found? Offer a hypothesis about why you obtained these results.
4. Were some organisms found only in certain places, or were all organisms you identified found everywhere? Hypothesize why you obtained those results.

Critical Thinking Questions

1. If you calculated results for multiple samples, compare these results. Discuss their similarities and differences. Offer hypotheses to explain any differences in the results.
2. If possible, compare these results with the results of samples taken at other times of the year. Discuss their similarities and differences. Offer hypotheses to explain any differences in the results.

3. If you calculated results for multiple samples and calculated the average final index value for these results, discuss the extent to which you believe this value represents the stream water quality accurately.
4. What organisms would you like to find to "really" raise the index?
5. Do you believe this benthic macroinvertebrate pollution-tolerance index is an accurate reflection of water quality? Why or why not?

Keeping Your Journal

1. In your journal, write the names of the organisms found in the sample data in this activity. Put a check mark beside the ones you think you can recognize. When you go out to your field site, try to recognize organisms from the laminated identification sheet. Circle the ones you find, and add the names of new organisms.
2. Of the organisms used in this lesson, which do you like best? Why?

Name

STUDENT ACTIVITY 4.5

Completing a Riverine Habitat Survey

Purpose

To introduce the habitat survey as a tool to aid in the evaluation of the suitability of a local river or stream environment for supporting aquatic life, taking into account the effects of humans on the river or stream.

Background

Based on the description and definitions of habitat parameters presented in Student Information 4.3, you will now complete a habitat survey for your field-study site. The characteristics included in this habitat assessment have been adapted from the U.S. EPA classification scheme and the stream classification guidelines for the Wisconsin Department of Natural Resources. Many states have developed their own forms to match their own data needs. Some are qualitative, like this one. Others are more quantitative, such as one used by the Illinois Department of Natural Resources. Each provides valuable information. If you are using a different assessment tool, your teacher may have you follow somewhat different procedures from those included here.

Your riverine habitat survey will allow you to organize a baseline of information on the river or stream you are studying. If other classes (or organizations) have already performed a habitat survey of your local river or stream, you will want to compare your results with theirs. When future classes complete the survey, they will be able to compare their observations with the ones you have made. For most favorable results, surveys

Materials

Per student

- life jacket and tow line (for deeper-water sites)
- rubber waders or boots (for shallower-water sites)
- journal

Per group

- Student Information 4.3: Riverine Habitats
- laminated topographic map of study site, 7.5-minute or 15-minute series
- USGS pamphlet, *Topographic Map Symbols*
- measuring tape, 50–100 meters
- Celsius thermometer, alcohol-filled with a metal jacket, or electronic; or temperature probe
- 25–50 cm of light cord
- stopwatch
- clipboard or firm writing surface
- blank paper
- completed data tables for Student Activity 4.4: Determining the Benthic Macroinvertebrate Index for Your River or Stream

Per class

- first-aid kit
- bottled water for washing
- soap and towels
- large plastic bags for collecting waste

Optional

- other material for determining site location
- camera, digital if possible
- compass

should be performed at the same time of year. If any conditions have changed, try to discover the reason for the change.

Investigators can also use good habitat assessments to pinpoint sources of water-quality problems by comparing test results and assessments of two sites on the same river or stream. For instance, if an upstream location yielded a better water-quality result, the habitat survey might give clues about what has affected the water between that site and yours. If test results show that your site has higher turbidity than a site upstream, you might notice that the assessments show your site's stream banks are less vegetated than those at the upstream site; construction just upriver from your site has also reduced the amount of vegetation in the watershed. If you had not noted both conditions on the habitat assessment, you would have fewer clues to work with when identifying the source of the problem.

Safety and Waste Disposal

Follow all field safety procedures presented in Lesson 1.

Procedure

1. Look over the riverine habitat survey before beginning your observations. While doing the survey, add supplementary notes to specific responses wherever appropriate.
2. Walk along the river or stream to the spot designated by your instructor.
3. Record river or stream name, time, date, and location. (Use the location identification technique your teacher indicates.)
4. Record current weather and water conditions, and note weather conditions for previous days, to the extent you are aware of them. Record your school and investigators.
5. Add any other significant location information or description or other visual observations of the water.
6. Each person in the group should make a sketch of the study section. If possible, take photographs.
7. On your sketch, use arrows to show the direction of the current in the stream.
8. Use a compass to determine which direction is north. Mark this on your sketch. Mark rapids, riffles, runs, pools, ditches, wetlands, dams, tributaries, landscape features, vegetation, and roads. Also note any outside features that may influence the stream (such as farm fields).
9. Obtain the temperature for the air and for the water in degrees Celsius, and enter on your habitat survey.
10. If your teacher provides instructions, determine the stream flow rate in meters per second. Enter your results on your habitat survey.
11. If your teacher provides instructions, determine the channel width in meters. Enter your results in your habitat survey.
12. Observe the stream channel of your river or stream at your site to respond to items 3–7 on your habitat survey.
13. Observe the substrate of your river or stream at your site to respond to items 8–12 on your habitat survey.
14. Observe the stream bank on both sides of your river or stream looking upstream from your site to respond to items 13–15 on your habitat survey.
15. Observe the vegetation at and near your river or stream at your site to respond to items 16–17 on your habitat survey.
16. Observe the local watershed that you can see from your site and on your topographic map to respond to items 18–19 on your habitat survey.
17. Note the odor and appearance of the water, using items 20 and 21 on the habitat survey.
18. Observe any animals in or around the river or stream, and observe any potential fish barriers. Enter your observations in items 22–26 on your habitat survey.
19. Observe any aquatic plants, including algae, in your field-study site. Enter your observations in items 27–32 on your habitat survey.
20. Observe the substrate and macroinvertebrates in or around the river or stream. Enter your observations in items 33–35 on your habitat survey.
21. Enter any additional relevant observations on your habitat survey.

Analyses and Conclusions

1. Write a few paragraphs summarizing the condition of your riverine habitat.
2. Based on your habitat survey, in what ways do you think the local watershed conditions affect the river or stream water? Which factors do you think have the greatest impact on your river or stream, especially on the aquatic animal forms?
3. Based on your habitat observations, evaluate the naturalness of this site.
4. Compare the visual biological survey, especially fish, with the rest of the survey. Discuss the degree of consistency or inconsistency.
5. What is the most damaging human factor to the water quality of your stream?
6. List five negative observations that you made. Rank them from the easiest to improve to most difficult to improve. Offer suggestions about steps your class could take to improve any of these undesirable situations.

Critical Thinking Questions

1. Based on your experience doing the survey, would you like your local government or other public entities to become more involved in your local river or stream? If so, describe the options they might look at in order to make an improvement.
2. Imagine that a new bridge is planned for this river or stream. Use the sketch you made, the topographic map if available, and your habitat survey results to determine which site for this bridge would have the least negative environmental impact. Hypothesize about whether the new structure would impact any critical life or important sites. Consider the construction process, not just the completed bridge.
3. How do you think this area would have looked a hundred years ago? Identify what you think would have been different and what might have been the same? What do you think will be different and the same in a hundred years?

Keeping Your Journal

1. What kind of organism would you most like to be in this habitat? Why?
2. Use drawing or writing to re-create this environment as it was before the influx of European settlers.
3. Using the habitat survey as a basis, prepare a children's book that can be used in an elementary school to teach about this river or stream.

Riverine Habitat Survey

River or Stream ________________ Location ______________________________

Date __________________________ Time _________________________________

Weather Conditions ___________ Weather Previous 24 Hrs ______________

Water Conditions _____________

School ________________________ Investigators _________________________

Other Location Information or Description

____

General Visual Observation of Current Water Conditions (Such as Water Level, Appearance)

____

Water Characteristics

1. Temperature
   water: ____ °C air: ____ °C

2. Stream velocity: _________ m/sec

Stream Channel Characteristics

3. Approximate channel width: ______ meters

4. Circle shape of the channel:
   a. narrow/deep c. wide/deep
   b. narrow/shallow d. wide/narrow

5. Circle approximate depth of runs:
   a. < 30 cm b. 30–60 cm c. > 60 cm

6. Circle approximate depth of pools:
   a. < 30 cm b. 30–60 cm c. > 60 cm

7. Looking upstream from the site, examine the stream channel for the following conditions. Mark 0 if absent, 1 if present, or 2 if clearly impacting the stream.

left		right
____	mud/silt/sand in stream	____
____	stream modified artificially	____
____	garbage/junk in stream	____

Substrate Characteristics

8. Circle any of the following stream habitats that are present.
   a. riffle b. run c. pool

9. Circle the size of the particles in stream bottom.
   a. silt/clay/mud c. gravel 3 mm to 5 cm e. boulders > 30 cm
   b. sand <3 mm d. cobbles 5–30 cm f. solid bedrock

10. Circle presence of logs and debris.
   a. none b. occasional c. common

11. Circle presence of natural materials such as twigs, grass, or leaves.
   a. none b. occasional c. common

12. Circle embeddedness of any rocks, cobble, and boulders.
   a. somewhat/not embedded d. completely embedded
   b. halfway embedded e. in gravel on the bottom
   c. mostly embedded

Stream Bank and Watershed Conditions

13. Looking upstream from the site, mark the slope of the stream bank.

left		right
____	vertical/undercut	____
____	sloping >30 degrees	____
____	gradual or no slope (<30 degrees)	____

14. Circle the extent of artificial (human-made) bank cover.
   a. 0–25% c. 50–75%
   b. 25–50% d. 75–100%

15. Looking upstream from the site, examine the stream bank for the following conditions. Mark 0 if absent, 1 if present, or 2 if clearly impacting the stream.

left		right
____	plant cover degraded	____
____	banks collapsed/eroded	____
____	banks artificially modified	____
____	garbage/junk on stream bank	____
____	loam or sheen on bank	____

16. Circle amount of shading of the stream or river site.
   a. open d. mostly shade
   b. mostly open e. shade
   c. half shade

17. Describe the stream-side natural cover upstream. Mark 0 if absent, 1 if present, 2 if dominant, or 3 if common.

left	a. *At water's edge*	right
____	evergreen trees	____
____	hardwood trees	____
____	bushes and shrubs	____
____	tall grasses, ferns, and so forth	____
____	lawn	____
____	boulders and rocks	____
____	gravel or sand	____
____	bare soil	____
____	pavement/structures	____

left	b. *Back-side to 10 meters*	right
____	evergreen trees	____
____	hardwood trees	____
____	bushes and shrubs	____
____	tall grasses, ferns, and so forth	____
____	lawn	____
____	boulders and rocks	____
____	gravel or sand	____
____	bare soil	____
____	pavement/structures	____

Cultural Factors

18. Look at land uses in the local watershed within 0.5 kilometers (¼ mile) upstream of the site. For each type of land use, mark 0 if absent, 1 if present, or 2 if clearly having an adverse effect on the water.

Established buildings

a. ___ single-family housing
b. ___ multifamily housing
c. ___ lawns
d. ___ commercial/institutional
e. ___ light industry
f. ___ heavy industry

Roads

g. ___ paved roads or bridges
h. ___ unpaved roads

Construction underway on

i. ___ housing development
j. ___ commercial development
k. ___ light industry
l. ___ heavy industry
m. ___ road bridge construction/repair

Agricultural

n. ___ grazing land
o. ___ feeding lots or animal holding areas
p. ___ inactive agricultural land
q. ___ cropland

Other

r. ___ mining or gravel pits
s. ___ logging
t. ___ recreation

19. Looking upstream from the site, look for the following possible point-source pollution activities. Note any others you observe. Mark 0 if absent, 1 if present, or 2 if clearly impacting the stream.

left		right
____	plant cover degraded	____
____	yard waste on bank	____
____	livestock in stream	____
____	pipes actively discharging	____
____	pipes entering stream	____
____	ditches entering stream	____
____	people swimming in stream	____
____	other: ________________	____

Aesthetic Characteristics

20. Circle any odor the water has.
    a. sewage c. rotten egg e. none
    b. chlorine d. fishy f. other (describe)

21. Circle the best description of the water's appearance.
    a. clear d. light brown, turbid g. oily sheen
    b. milky e. dark brown h. greenish
    c. foamy f. reddish i. other (describe)

Visual Biological Survey

22. Types of wildlife observed in or around the site. (Note the names and number of specific species.)
    ____ amphibians ____ reptiles
    ____ waterfowl ____ mammals

23. Can you observe any fish in the river or stream?
    no yes

24. If fish observed, indicate number in each type of riverine habitat.
    deep pools ____ deep runs ____
    shallow pools ____ shallow runs ____
    riffles ____

25. If fish are present, indicate number in each size range.
    small (2–5 cm) ____ medium (5–15 cm) ____ large (16 cm and over) ____

26. Note any fish barriers visible.
    beaver dam ____ road barrier ____ waterfall ____
    other ____ dam ____ none ____

27. Note the extent of aquatic plants in the stream.
    none ____ occasional ____ plentiful ____

28. If aquatic plants are present, are they attached or free-floating?
    attached ____ free-floating ____

29. Note the location of aquatic plants.
    stream bank ____ pools ____ near riffle ____

30. Note the extent to which submerged stones, twigs, or other material in the stream are coated with a layer of algae "slime." Note also the extent to which clumps or maps of algae are floating in the water.
    none ____ occasional ____ plentiful ____

31. If algae are present in this form, note the extent of algae "slime" coating.
    light ____ heavy ____

32. If algae are present in this form, note the color of the algae "slime" coating.
    brownish ____ greenish ____ other (describe) ____

Macroinvertebrate Survey

33. If macroinvertebrates were collected from the stream bottom, which type of method was used for this habitat?
    ____ rock-scraping method, from cobbles and large stones selected from riffles.
    ____ stick-picking method, from woody objects in streams with sandy, silty bottoms.

34. Note the extent to which macroinvertebrates are present.
    none ____ occasional ____ plentiful ____

35. If present, note the types of macroinvertebrates found. Check as many as apply.

	occasional	plentiful
wormlike	____	____
snails/clamlike	____	____
insects	____	____
crayfish	____	____
other (describe)	____	____

Other remarks (describe):

Name

STUDENT ACTIVITY 4.6

Determining the Diversity Index For Your River or Stream

Purpose

To determine the diversity index for a field site at your local river or stream, as a measure of water quality.

Background

Species diversity is important to the riverine community, and it generally provides a good indicator of environmental health. The diversity index allows scientists to compare environments using a mathematical model of how the number of species and the number of individuals are apportioned and related to each other at a specific site. You can use the index to summarize a lot of data into one set of numbers. For the purposes of this diversity index, the separation of organisms in different taxa (apparent species) is needed, not the total identification of each organism.

In calculating the diversity index, you will use the diversity index equation. This equation allocates the appropriate weight to the numbers and diversity of taxa in the collection. The diversity index equation is:

$$H = -\sum_{i=1}^{s} P_i (\log_e P_i),$$

where
H = diversity index, and
P = proportion of taxa

H = inverse sign of (P) (log P)
P, the proportion of taxa, is obtained by dividing the number of organisms per taxon by the total number of organisms in the collection.

You will set up your data table to facilitate this set of calculations. Because some calculators do not compute logarithms, an alternate method of computing (P)(log P) is provided in the data table. (The table can be set up on a computer spreadsheet for rapid calculations and easy comparison of samples, with results also quickly available for pasting into other reports.)

The diversity value, H, reflects a combination of several community factors that indicate environmental health—richness, density, and evenness. The number of taxa in a sample, or **richness,** shows the apparent variety of species present. The second factor is the relative **density** of each taxon, the number of organisms for each group in one sample or area. Too many of one species is often a sign of pollution or environmental degradation. In polluted waters, some tolerant species are able to multiply rapidly, replacing a number of more sensitive species.

The third factor that contributes to diversity is **evenness,** the number of individuals in each taxon relative to the total number of individuals in the sample. In the diversity equation, evenness is expressed as P. Evenness shows how uniformly the individuals are

Materials

Per group

- completed data tables from Student Activity 4.4: Determining the Benthic Macroinvertebrate Index for Your River or Stream
- calculator with logarithmic function

For the teacher

- overhead transparency of Teacher Table 4-3, Model of Diversity Index

distributed among the taxa. In an even community, the P-values for each taxon will be similar.

The diversity index usually ranges from 1 to 3. A diversity index of 1 indicates that the organisms in the sample are from one species. Some factor, such as pollution or seasonal water temperature, is limiting the population. A diversity index of 3 or greater indicates many species, with relatively similar numbers in each taxon. First, you may practice producing a diversity index with data provided by your teacher. Then, after returning from the field, you will produce a diversity index using data you gathered as part of Student Activity 4.4: Determining the Benthic Macroinvertebrate Index for Your River or Stream.

Procedure

1. Transfer the data on date, time, location, weather, water conditions, and other site information from your Student Activity 4.4: Determining the Benthic Macroinvertebrate Index for Your River or Stream onto your Observations for this activity.
2. Transfer the data from Sample 1 on benthic macroinvertebrates at your field site from your Data Table 1, Survey of Benthic Macroinvertebrates, in Student Activity 4.4 to your own Data Table 1 for this activity, Diversity Index for Sample of Benthic Macroinvertebrates. Enter the name of each taxon in Column A, and the number of organisms for each taxon on the corresponding line in Column B.
3. Calculate and record the total number of taxa.
4. Calculate and record the total number of organisms.
5. Calculate and record the proportion of taxa (P) for each taxon. To do this, divide the number of species in that taxon by the total number of organisms.
6. Calculate and record (log P) for the value of P for each of the taxonomic groups. If your calculator computes logarithms, use this feature to compute (log P). If your computer does not compute logarithms, do the following steps for each value of P:

   Enter P into the calculator
   Press log 10
   Press division key
   Enter .301
   The result is (log P).
7. Calculate the index value for each taxon by multiplying P by (log P). Record the results.
8. Total the index values for all the taxa.
9. To find the diversity index (H), take the inverse sign of the total of the diversity index values determined in the preceding step. The diversity index will be an integer ranging from 0 to a number greater than 3. Enter the result in Data Table 1 and Data Table 2.
10. Repeat Steps 2–9 for each sample, using a separate Data Table 1.
11. Sum the diversity index for all samples. Enter your results in your own Data Table 2. Divide that total by the number of samples. Record the result as the average diversity index.

Observations

River or Stream ______________ Location ______________

Date ______________ Time ______________

Weather Conditions ______________ Weather Previous 24 hrs ______________

Water Conditions ______________ Air Temp ____°C Flow Rate ____ m/sec

School ______________ Investigators ______________

DATA TABLE 1:

Diversity Index for Sample of Benthic Marcroinvertebrates

Sample Number: ______

Taxon (A)	Number of Organisms (B)	P (Orgs in Taxon / Total Orgs)	P (decimal form)	P Log 10	Log P (P Log 10 / .301)	Index Value (Log P•P)
(make one row for each taxon)	____					____
Total	____					____
Diversity index (inverse of total of index values)						____

DATA TABLE 2: Average Diversity Index

Diversity Index for Sample 1 = __________

Diversity Index for Sample 2 = __________

Diversity Index for Sample 3 = __________

Average Diversity Index = __________

Analyses and Conclusions

Diversity Index Water-Quality Indications

Diversity Index	Water Quality
<1	Few taxa, some with many individuals; may indicate heavily polluted water
1–3	May indicate moderately polluted water
3+	May indicate relatively clean or unpolluted water

1. Using the diversity index for your river or stream, and the preceding chart, assess how polluted your river or stream is likely to be.
2. Would you consider your sample rich in species composition? Why? How does the richness of the sample affect the diversity index?
3. How did the density of the taxa differ? What do you think may cause some taxa to be more represented in your sample data?
4. Were there equal numbers of each taxon? What is this measure called?

Critical Thinking Questions

1. Compare the index values for the diversity index and with the benthic macroinvertebrate pollution-tolerance index obtained for your river or stream in Student Activity 4.4. How are they consistent or inconsistent? Hypothesize why that is so.
2. Do you believe this diversity index is an accurate reflection of the water quality of your river or stream? Why or why not?

Keeping Your Journal

1. Compare the actual results of this index with the results you predicted earlier in *Rivers Biology*. Account for differences.
2. Compare how you feel about these results with how you expected to feel about them.

Name

A River Diorama

Purpose

To use the information learned about the riverine environment to create a diorama of that environment.

Introduction

In this assessment, you will create a diorama of a river in order to demonstrate what you have learned about rivers. Your river diorama should contain the elements found in the habitat assessment, as well as physical parameters, animals, and plants.

Procedure

1. In groups or as individuals, as directed by your teacher, choose a portion of the riverine environment for which you wish to construct a diorama.
2. You or your group should brainstorm a list of all the animals and plants that might live in or near your river.
3. Make a complete list of the animals and plants in your chosen environment. Find as many pictures as you can of them. Draw those that are not available. If you have photocopies, color them. Keep scale in mind. Carp, for example, should be bigger than mayflies.
4. Use the cardboard box, paper, and other art materials to create a diorama. Place or paint the substrate, river, banks, large trees, and other elements of your riverine environment.
5. Place the appropriate animal and plant pictures in the appropriate locations in the diorama. Label as needed.
6. Place your diorama on display as directed by your teacher, with a label that identifies those who created it.
7. Prepare a paper to summarize this habitat and indicate the species of organisms that are present in context. In this paper, you should explain the drawings and pictures of animals and plants in a discussion of the environment being represented. For instance, you may want to show a riffle area where the class collected. It may have rock, critters, and tree limbs shown in the water. You might describe the drawing by giving the names of these animals and how they relate to each other and to the water.
8. In your paper, describe the water quality of your chosen environment in terms of the organisms found. Use any indices as you rate and describe the habitat and its water quality.

Materials

Per group

- pictures or photocopies of animals found in your river
- pictures or photocopies of plants found in your river
- markers or colored pencils
- cardboard box

Per class

- poster board
- finger or tempera paint
- brushes
- miscellaneous art materials
- matte knife or strong scissors
- computer and printer

Performance Criteria

- Feature a single riverine habitat throughout the diorama. The location is evident through pictures and organization and is stated in the written paper.
- Include in the diorama the plants and animals appropriate to that single riverine habitat (avoid inappropriate organisms).
- Clearly demonstrate organization and planning. The diorama is neatly constructed.
- Provide an accompanying paper that explains the habitat depicted, and the role of the depicted animals and plants shown in that habitat.
- Use and explain indices in the paper to indicate water quality.
- Use correct grammar, sentence structure, and spelling in the paper.

LESSON 5 Dissolved Oxygen in River and Stream Water

Focus

The purpose of this lesson is twofold: to introduce water oxygenation and oxygen levels as processes essential for aquatic organisms and to give students introductory experiences in conducting a dissolved oxygen test.

Learner Outcomes

Students will:

1. Learn the meaning of dissolved oxygen (DO) and biochemical oxygen demand (BOD).
2. Understand why fish and other aquatic life need dissolved oxygen.
3. Learn what factors, including nutrients and human factors, may affect the amount of dissolved oxygen in a waterway.
4. Conduct tests of dissolved oxygen and biological oxygen demands in a river or stream, and use their data to determine water quality.

Time

Four class periods of 40–50 minutes per period and a field trip to a river or stream

DAY 1: Student Information 5.1: Oxygen in Water

DAY 2: Student Information 5.2: What Is Biochemical Oxygen Demand?

DAY 3: Student Activity 5.3: Dissolving Oxygen in Water

FIELD TRIP: Student Activity 5.4: Measuring Dissolved Oxygen in a River or Stream (Parts A and B at field site, Part C in lab)

DAY 4: (five days after field trip) Student Activity 5.5: Measuring Biochemical Oxygen Demand in River or Stream Water

Safety and Waste Disposal

Follow all lab and field safety procedures presented in Lesson 1. Chemicals used in student activities **cannot** be flushed down the sink. Determining dissolved-oxygen concentrations involves use of caustic materials, so follow the test-kit manufacturer's procedures and precautions carefully. Have students place all liquid wastes from these activities in a waste container labeled "heavy metals." Evaporate the liquids under a hood in the classroom. A used gallon milk container properly labeled works well as a waste container. When the container is full, or once a year, place the remaining solid residue in a plastic bag for disposal in a landfill. Place solid waste in plastic garbage bags. For more on waste handling, see *Rivers Chemistry*.

Advance Preparation

Prepare to supply students with Student Information and Activity sheets 5.1 through 5.5. Gather all necessary equipment and materials. Become familiar with the specific directions provided by the manufacturer of the dissolved oxygen kit your students will be using. Practice the dissolved-oxygen testing procedure described in Student Information 5.1, Student Activity 5.3, and Background for the Teacher so you will become familiar with the technique and can model a procedure that avoids problems. If desired, prepare a short lecture on dissolved oxygen in water and its relationship to aquatic life.

Prepare the dissolved-oxygen test kits. Clean and acid-wash the DO bottles. If you wish students to do the acid wash, see *Rivers Chemistry* for specific teaching activities on this skill. Prepare and label waste-collection containers. Laminate copies of the instructions for each DO kit.

As part of this lesson, each group will collect a sample of river water in a DO bottle for the test of biochemical oxygen demand. Test kits generally supply only one DO bottle; if necessary, obtain additional DO bottles. (Each group will use one bottle for the DO test, then use that bottle to collect a sample for the BOD test.) If desired, obtain more dissolved-oxygen bottles so you can instruct your students to collect additional samples.

Review water-sampling equipment and techniques, especially safety procedures. Review information in the unit introduction on field trip management. After water collected in Student Activity 5.4 has incubated for five days, students will perform Student Activity 5.5. Therefore, make certain the fifth day does not fall on a day that school is not in session. If you plan the field trip for Student Activity 5.4 for a Thursday or Friday, Student Activity 5.5 can take place the following Tuesday or Wednesday.

Materials

Student Activity 5.3: Dissolving Oxygen in Water

Per student

safety goggles, lab apron, and gloves

Per group

1L water from nearby pond or stream

1L dechlorinated tap water

wide-mouth container, at least 1-L (such as glass jar, battery jar, or milk carton with top cut off)

dissolved-oxygen bottle with stopper, 60-mL

dissolved-oxygen test kit

laminated instructions for the dissolved-oxygen test kit

paper towels

Celsius thermometer

aluminum foil, 30 cm (8 in.)

stirring rod or spoon

200 mL deionized or distilled water, if running control

10 ice cubes

Per class
freshwater aquarium with aerator pump
Elodea or similar aquatic plant, in container of chlorine-free water
egg beater
forceps
waste container, 1–2 L
first-aid kit
Optional
additional models of aquatic environments, such as a polluted aquarium with no animals or an aquarium with many fish
light source

Student Activity 5.4: Measuring Dissolved Oxygen in a River or Stream
Per student
safety goggles, lab apron, and gloves
life jacket and tow line (for deeper-water sites)
rubber waders or boots (for shallower-water sites)
Student Activity 5.3: Dissolving Oxygen in Water
journal
Per group
Celsius thermometer, alcohol-filled with a metal jacket, or electronic
dissolved-oxygen test kit
dissolved-oxygen bottle with stopper
laminated instructions for the dissolved-oxygen test kit
water-sampling pole, with clamp
aluminum foil, 30 cm (8 in.)
labeling tape or marker
plastic garbage bag (for trash)
straightedge or ruler
Per class
topographic map or other materials for determining location
deionized or distilled water, about 2 L
cooler, with ice, for transporting DO bottles
waste container, 1–2 L
lightproof incubator set at 20°C, or lab drawer with Celsius thermometer

Student Activity 5.5: Measuring Biochemical Oxygen Demand in River or Stream Water
Per student
safety goggles, lab apron, and gloves
calculator (optional)
Per group
water-filled DO bottle from Student Activity 5.4, incubated for five days
dissolved-oxygen test kit
laminated instructions for the dissolved-oxygen test kit
waste container, 1–2 L
Per class
lightproof incubator set at 20°C, or lab drawer with Celsius thermometer

Vocabulary

aerobic organism
aerobic oxidation
algal bloom
anaerobic bacteria
bioassay
biochemical oxygen demand (BOD)
control
dissolved oxygen (DO)
Elodea
eutrophication
fish kill
metabolism
percent saturation
titration
Winkler method

Background for the Teacher

In order to add the chemicals during a dissolved-oxygen test, the user must remove the stopper from the bottle, add the chemicals, then replace the stopper. Students may have difficulty keeping air bubbles out of the bottle while performing this procedure. They can minimize this problem by placing the stopper in the bottle with moderate speed and gentleness. If the user inserts the stopper too slow or fast, air bubbles have a greater tendency to form.

If a tiny air bubble is present before the oxygen is fixed, you may place small glass beads or small lengths of glass tubing into the bottle, so no water splashes out; as the liquid rises, the small air bubble will be expelled. Continue the test. This method may produce slightly inaccurate results; if this is used on only one of three samples, averaging the results minimizes the inaccuracy.

For most classes, Student Activity 5.4: Measuring Dissolved Oxygen in a River or Stream is the first field activity in which students use prepackaged solid chemicals at the field site. Students will discover that doing labwork at the site differs from working in a school lab where they have access to tables and other conveniences. Opening chemical packets while the wind is blowing, and completing a test while sitting on a river bank, can be difficult. Remind students to pick up all paper and plastic containers. If your site has less than perfect conditions, advise your students to take extra caution.

Plan the timing and location of the field trip for Student Activity 5.4 so the river or stream from which they take the sample for Student Activity 5.5 is between 15°C and 20°C. The greater the temperature difference between the water source and the incubator, the more inaccurate the test. Take care to keep the incubating water samples at 20°C or less. If the temperature of the sample exceeds 20°C, dissolved oxygen will come out of solution. Prevent this by placing the bottles in ice to get them back to the classroom. Putting the ground glass stopper in the DO bottle firmly should form a tight seal, creating a closed system, preventing dissolved oxygen from escaping. If the system is not closed, dissolved oxygen will escape, yielding a high BOD value even if no organic matter is present.

Introducing the Lesson

1. If not already done, assign reading of Student Information 5.1: Oxygen in Water as homework.
2. Have students discuss and answer the questions for Student Information 5.1. (Answers for student sheets are in Appendix B.) (Some of the material in this handout is also presented in *Rivers Chemistry*.) Discuss how oxygen levels determine what organisms can be found in a body of water. Discuss how seasonal temperatures and pollution may affect the amount of dissolved oxygen in a river or stream.
3. Have students read, discuss, and answer the questions for Student Information 5.2: What Is Biochemical Oxygen Demand?. (Some of the material in this handout is also presented in *Rivers Chemistry*.)
4. Discuss human factors that can affect the temperature of a waterway. Make sure students understand how seasonal temperatures affect the amount of dissolved oxygen in a river or stream. If desired, have students hypothesize about the dissolved oxygen and biochemical oxygen demand status of their field site, including speculation about what factors influence those results.

Developing the Lesson

1. Have students read Student Activity 5.3: Dissolving Oxygen in Water. Discuss possible water sampling sources in the classroom, such as distilled water, pond water, tap water, and aquarium water. You can tape a letter on each source or write the name of the water source. (With long names, a letter might be easier to write on the DO bottles.) As a control, one group can use distilled water, or have all groups use distilled water to check their technique. Also discuss ways of increasing the oxygen in a sample. As described in this activity, preview proper procedures for obtaining the water sample in the DO bottles. You may wish to demonstrate good technique.
2. In groups, have students carry out Student Activity 5.3: Dissolving Oxygen in Water, with each group testing at least one different water type, aeration technique, or both. If you want students to calculate and consider the standard deviation of results of multiple tests, see specific teaching materials in *Rivers Chemistry*. (This process is not necessary for *Rivers Biology* as presented here.) Discuss results and answers to questions.
3. As students carry out the activities in Lesson 5, observe and evaluate their individual performance on laboratory skills, recording your assessment on the Biological Field-Study Proficiency Checklist in Appendix C, Assessment Tools.
4. At the field site, have students carry out Student Activity 5.4: Measuring Dissolved Oxygen in a River or Stream, Part A (dissolved oxygen testing) and Part B (collecting samples for the test of biochemical oxygen demand). (This material is adapted from *Rivers Chemistry*.) Each group

can use the DO bottle in which they later collect the sample for the BOD test. Indicate what technique students should use for determining location. Decide how many tests each groups should run. You may, for instance, have students do two tests if they are new to DO testing. If these tests produce the same results, then students are using good technique. If they differ, have students run another test. Experienced groups can conduct one test and, during analysis, compare results with those of other groups. (If you have sufficient DO bottles, instruct each group to collect several samples.)

5. In the classroom, have students complete Student Activity 5.4, including Part C, placing the samples in an incubator.

Concluding the Lesson

1. Five days after the field trip, have students do Student Activity 5.5: Measuring Biochemical Oxygen Demand in River or Stream Water. (This material was originally prepared for *Rivers Chemistry*.)
2. Discuss with students ways to keep dissolved oxygen levels at a comfortable level for fish and other aquatic life. Consider seasonal changes, temperature, rainfall, and pollution from surrounding farms, industries, and residential developments.

Assessing the Lesson

1. Review and assess student performance on the laboratory skills demonstrated during this lesson, as recorded on Biological Field-Study Proficiency Checklist, in Appendix A, Assessment Tools.
2. Have students add to their river or stream collages.

Extending the Lesson

1. Have students contact a local fishery or wildlife agency to find out how much dissolved oxygen various species of fish require. For specific teaching activities on doing nonfiction research, see *Rivers Language Arts.*
2. Invite a fishery or wildlife professional to visit the class so students can ask questions regarding dissolved oxygen and other topics of student interest.
3. Have students write a short story (perhaps a children's story) about how a fish might react if its supply of dissolved oxygen were suddenly reduced or depleted. For specific teaching activities on creative writing, see *Rivers Language Arts.*
4. Have students research the environment upstream from your field-study site to determine if it includes farms, paper mills, food-processing plants, or other human sources that are adding organic matter to the river or stream.
5. Visit a local wastewater (or sewage) treatment plant to see how wastewater is treated before release to the environment. Alternatively, invite an official from the plant to talk with your class about wastewater treatment procedures.
6. Visit a local farm or industrial plant to observe controls on organic pollution and discuss state and federal requirements for handling waste materials. Obtain permission for students to interview farm or plant officials to determine how much organic matter is being released into the local environment. Before the field trip, provide students with specific instructions on how to prepare a written or oral report that summarizes their observations and conclusions. For specific teaching activities on writing scientific reports, see *Rivers Language Arts.*

Name

S T U D E N T
INFORMATION
5.1

Oxygen in Water

The single most important determinant for the presence of aquatic organisms in a river, stream, or lake is the amount of oxygen its water contains. The chemical formula for water, H_2O, shows that oxygen atoms are always present in water molecules. The oxygen in water is not available for organisms to breathe, however, because it is tied up in this chemical bond.

For respiration, most aquatic life needs oxygen in the free elemental state (O_2). Because air is approximately 21 percent oxygen and 78 percent nitrogen, this elemental oxygen is readily available in the atmosphere. Just as you need an adequate amount of oxygen gas in the air to live, so fish and other aquatic life need adequate amounts of oxygen gas dissolved in water.

Just as many solids will dissolve in liquids, so will gases. For instance, carbonated drinks consist of carbon dioxide gas dissolved in water, juice, or some other liquid. Most natural bodies of water contain dissolved oxygen. **Dissolved oxygen** is elemental oxygen (O_2) gas in a solution of water. Temperature, green plants, and mixing all influence the amount of oxygen gas dissolved in a particular river or stream.

How Does Temperature Impact Oxygen Levels?

Temperature influences the amount of oxygen gas in river or stream water. Gases generally dissolve better in cold water than in hot water. Oxygen molecules move more slowly in cold water than in warm water, so the oxygen and other gases do not escape as quickly. So dissolved oxygen levels may vary dramatically throughout the year or day. More oxygen is usually dissolved in a river or stream during winter months than during summer. More oxygen can be dissolved in water at night or during early morning than in the late afternoon after the water has been warmed by the sun.

In deep rivers, more oxygen gas can dissolve in the colder water at the bottom than in the warmer water near the top. At lower depths, water is also under greater pressure, which further increases the amount of dissolved oxygen that water can contain.

So, when measuring dissolved oxygen, always measure the temperature of the sample. When you measure dissolved oxygen in *Rivers Biology,* you will use an oxygen percent saturation chart in which the temperature effect is taken into account.

Water Temperature and Aquatic Organisms

Water temperature does not affect all types of aquatic organisms equally, both in terms of dissolved oxygen content and the actual temperature of the water. Temperature ranges greater than 20°C (68°F) are considered warm waters for the health of aquatic organisms. Temperatures at this level support considerable plant life. They can, however, also support many fish diseases in bass, crappie, bluegill, carp, catfish, and their relatives. Hot water stresses many colder-water organisms. Warm waters support the more tolerant, less-sensitive benthic macroinvertebrates.

Middle-range temperatures of 12°C to 20°C (55°F to 68°F) support less plant life but a greater diversity of fishes. At these temperatures, northern pike, salmon, trout, and walleye thrive, and fewer fish diseases are found. Moderately tolerant benthic macroinvertebrates such as mayflies, stone flies, and water beetles occur in more abundance.

Streams at less than 12°C (55°F) support far fewer varieties of life. They have almost no plants, and trout is a major vertebrate. Benthic macroinvertebrates such as stoneflies, caddisflies, dobsonflies, and mayflies are abundant and diverse.

How Do Aquatic Plants Impact Dissolved Oxygen Levels?

If temperature affects the potential of the water to retain oxygen gas, how does the water actually acquire oxygen? One of the main ways slow-moving water obtains oxygen is through photosynthesis of green aquatic plants and phytoplankton. Such organisms produce oxygen during sunny days as a by product of photosynthesis and deplete it during the night in respiration. So, the level of dissolved oxygen in slow-moving water can depend on how much algae and green plants it contains.

Aquatic plants use oxygen, but, during photosynthesis, they produce more oxygen than they use. Slow streams with many plants may show daily variation in oxygen levels. On a warm summer day, significant photosynthetic activity may drive oxygen levels quite high by late afternoon. Early-morning oxygen levels may drop drastically as plants and animals use up oxygen via respiration. Dissolved oxygen levels are normally lower in the morning, due to the respiration activities of aerobic organisms.

High water temperatures in late summer and early fall increase abnormally rapid algae growth, called **algal** (AL juhl) **bloom.** An increase in algae populations leads to a corresponding increase in aerobic-decomposing bacteria. This, in turn, can lead to rapidly declining oxygen levels, resulting in a summer **fish kill,** the death of large numbers of fish from insufficient oxygen levels or toxic chemicals in a waterway.

How Does Oxygen Mix With Water?

Oxygenation of water also takes place by mixing water and air. Wherever air interfaces with water, atmospheric oxygen dissolves in the water. Streams and rivers with high oxygen levels are usually turbulent, with water flowing over falls and rocks (or moving through turbines). As the water splashes into the air, the two substances mix, causing the oxygen gas in the air to dissolve into the water. Artificial aquatic environments, such as home fishtanks, use pumps to mix air with the water.

DRIFTWOOD

"When you're conserving a river, you're conserving a life."

Kevin Coyle
American Rivers

How Does Human Activity Affect Dissolved Oxygen Levels?

Humans engage in industrial, agricultural, commercial, and residential activities that produce waste that may enter the river or stream through wastewater (sewers and storm drains), runoff, accidents, or deliberate dumping. These activities may yield chemical (inorganic), animal, human, or other organic waste.

Oxygen is very active chemically, meaning it can combine with a wide range of chemicals or react spontaneously with organic compounds. So, if chemical pollutants are present through human waste, great quantities of the oxygen in the river or stream will combine with those chemicals, thus lowering the level of dissolved oxygen.

When excess organic material (whether human, animal, or plant) enters the water, the decomposition of those materials by aerobic bacteria depletes the water of oxygen. Many of these organic materials serve as nutrients, which result in algae bloom. This, in turn, causes a disproportionate population of aerobically respiring bacteria which further leads to declining levels of dissolved oxygen.

Human activities may also cause erosion, which increases sediment in the water. Sediments normally enter waterways as a result of runoff during heavy rains. In flowing waters, they often remain suspended in the water, reducing light penetration and inhibiting photosynthesis and subsequent oxygen production. Sediments may also contain nutrients that, when released into the water, encourage algae growth and the further decline of oxygen.

Power plants and industries that use river water to remove their waste thermal energy can also have a major effect on the amount of oxygen dissolved in the water. When the industrial operations return the water to its river source, it may be 10°C (or more) hotter than the upstream water. This warmth may increase the number of certain aquatic plants by extending their growing season. Fish and other aquatic life, however, may die because of lowered levels of dissolved oxygen in the warm water.

How Do Organisms Use Oxygen?

Oxygen is a limiting factor for many aquatic forms, but not all aquatic organisms require the same dissolved-oxygen level. Generally, aquatic vertebrates and most other aquatic life are stressed when the oxygen concentration is 5.0 milligrams of oxygen per liter (mg/L), also expressed as parts per million (ppm), or less.

Most fish require a minimum of 4.0 mg/L just to survive over an extended time period. Some Mississippi River fish, such as buffalo, carp, and catfish, require only 4 to 8 mg/L. Species that require much higher oxygen concentrations (6 mg/L) live in cold water. Trout and salmon need 8 to 15 mg/L of dissolved oxygen, so they perish in waters where catfish would thrive.

Various life forms have different ways of extracting oxygen from the air or water in which they live. Some animals, such as earthworms and tadpoles, obtain much of their oxygen by diffusion through their skin from water or air. Others have evolved lungs and gills. Most amphibians go through a metamorphosis that requires three components: gills, lungs, and diffusion. Generally, the more evolutionarily advanced the organism, the more complex its procedure for removing and using the oxygen.

Organisms that require oxygen are called **aerobic organisms** (er OH bik OR guh niz uhms). Oxygen can be toxic to other life forms, such as **anaerobic bacteria** (an er OH bik bak TEER ee ah), bacteria that do not require oxygen. Many anaerobic organisms cannot survive in high oxygen levels, because it tends to impede enzymatic activities necessary for anaerobic respiration.

The Effect of Low Dissolved-Oxygen Levels

If oxygen levels drop, some aquatic species may leave or die; others may increase in number. Larger animals, such as fish and amphibians, may move away from a locally affected area. Benthic macroorganisms, which spend their entire lives in a one-square-meter world, are subject to the conditions at that site. When oxygen consumption exceeds the supply of oxygen, anaerobic conditions exist. For the many of these organisms that are aerobic, low levels of oxygen prevent normal metabolic activities, threatening survival.

When temperature, chemical processes, or organic decomposition result in abnormal algae growth and the corresponding increase in aerobic forms of life, a rapid decline of oxygen occurs in an otherwise healthy body of water and that body of water becomes unable to support oxygen-dependent organisms. This process, called **eutrophication,** causes the death of that body of river, stream, lake, or pond.

Why Does Some Mud Smell So Bad?

Have you ever noticed that warm, stagnant ponds, especially the mud, smell bad, while cold mountain streams do not? Bad smells and low oxygen levels are generally correlated.

When all or nearly all of the dissolved oxygen is depleted from a body or water, the oxygen-dependent organisms in it die; only anaerobic organisms survive.

Some anaerobic bacteria obtain oxygen by reducing sulfates in the water to hydrogen sulfide. Hydrogen sulfide gas is very toxic and has a distinctive and objectionable odor of rotten eggs. Even small amounts of hydrogen sulfide (less than 1 ppm) can give a water supply a very bad taste and odor.

Decomposition of waste organic material requires oxygen, so if the water has no dissolved oxygen, decomposition cannot continue. The incomplete, partial oxidation of organic wastes produces a variety of organic compounds, including alcohols, aldehydes, carboxylic acids, and ketones. They, along with the hydrogen sulfide, yield a dark, foul-smelling mixture sometimes found in stagnant areas. The foul smell and dark color confirm that the water contains no dissolved oxygen.

How Is Dissolved Oxygen Measured?

Many scientific suppliers make dissolved-oxygen (DO) kits for class use, but they all work in a similar way. First, you add a chemical, which reacts with the oxygen that is dissolved in the water in a special DO bottle. This chemical reaction forms an insoluble colored precipitate. At this point, the oxygen is "fixed."

Up through this step, the DO bottle must have no air bubbles. After the oxygen has been fixed, air bubbles in the bottle are not as crucial, although it is still good laboratory technique to have no air bubbles. If air bubbles are present before the oxygen is fixed, abort the test and start over with a new sample.

Then you will add other chemicals drop by drop (**titration**) until the solution becomes colorless, an indicator that the first chemical is completely dissolved. To determine the amount of dissolved oxygen (in mg/L), you will count the number of drops of chemical added in the final step. Chemists refer to this test as a modified **Winkler method.**

When completing a chemical test, good procedure often includes checking the viability of the chemicals you will be using. Scientists do this by running the test on a substance whose characteristics are already known, called a **control.** For instance, a control for testing oxygen in water would be to test water that contains no oxygen. Distilled water, because it has been heated, has no oxygen. A test of distilled water for dissolved oxygen should detect no oxygen.

Questions

1. In your own words, what is dissolved oxygen, and why is it important in aquatic habitats?
2. How does the solubility of gases in liquids change with shifts in temperature?
3. From what you know about the kinds of fish found in your river or stream, what temperature, dissolved oxygen, and percent saturation values do you expect to find?
4. Do you think a different kind of fish might become common? Explain your reasoning.
5. One summer morning, some teenagers stopped at a farm pond. They noticed that the top of the water was covered with dead fish. Propose a hypothesis for why this happened. What test results would you expect to find if you tested that pond for dissolved oxygen?
6. Marita and Jose went fishing one hot summer day. Marita let her fishing line hang several meters deeper in the river than Jose. She caught her limit of fish, while he caught only one catfish. Offer a hypothesis as to why Marita caught more fish than Joe. How would you test to prove that hypothesis?

DRIFTWOOD

"If there is magic on this planet, it is contained in water."

Loren Eiseley

Name

STUDENT INFORMATION

5.2

What Is Biochemical Oxygen Demand?

Even as photosynthesis and aeration are adding oxygen to water, organic matter is removing it. The microorganisms in a river or stream feed on the organic matter, and the process by which they break down that organic matter reduces the supply of dissolved oxygen gas in the water. The chemical process, known as **aerobic oxidation** (er OH bik AHX ih day shun), occurs when bacteria use oxygen to convert organic matter into simple energy-rich compounds and inorganic substances, such as carbon dioxide and water. The greater this rate of conversion, the greater the threat to aquatic organisms of insufficient or low levels of dissolved oxygen.

All rivers and streams naturally contain some organic matter produced by aquatic organisms. Leaves, sticks, fruit matter, and body wastes from animals wash into streams and rivers from swamps, fields, forests, and other environments along river banks. These natural activities act as a food source for stream inhabitants. Human activities, however, can lead to excess organic matter in a river or stream. When this happens, the populations of microorganisms that feed on such matter can explode, consuming greater amounts of dissolved oxygen.

Aerobic microorganisms use dissolved oxygen in the water to break down organic matter and for other activities of **metabolism,** all the chemical activities that take place in an organism. If the amount of organic matter and the number of microorganisms become too great, these activities may deplete almost all the oxygen in the water, killing oxygen-breathing aquatic species. The requirement, or demand, for oxygen that organic and chemical matter places on water is called **biochemical oxygen demand,** or **BOD.** BOD is an indirect measurement of how much stress is being placed on the dissolved-oxygen system of a waterway. The lower the stress, the more oxygen is available for aquatic organisms such as fish and shellfish.

Human Causes of Increased BOD

In some rivers and streams, humans overload the natural system by dumping additional organic materials. These sources may include pulp and paper mills, meatpacking plants, food-processing plants, wastewater treatment plants, urban and agricultural runoff, nutrients from lawn fertilizers, and leaves or grass clippings. Rain and melting snow can carry sewage from sanitary sewer connections into storm water drains. As was covered earlier in *Rivers Biology,* chemicals in some of these materials may combine with oxygen in the water to deplete available oxygen. Additionally, decomposition of such materials by aerobic bacteria may also cause such depletion.

Moreover, the ecosystem becomes dependent on extra nutrients contributed by these human sources. If this food source stops, the organisms die. When the organisms die, aerobic bacteria consume large amounts of dissolved oxygen in the decomposition process. Under extreme conditions, the amount of dissolved oxygen in the aquatic habitat may fall to near zero.

DRIFTWOOD

"Water is H_2O, hydrogen two parts, oxygen one, but there is also a third thing, that makes water and nobody knows what that is."

D. H. Lawrence, *Pansies*, 1929

How Do You Test for Biochemical Oxygen Demand?

Both DO and BOD tests are indicators of water quality. Biologists and other scientists test for BOD as an indicator of the quantity of oxygen used in the metabolism of microorganisms and in the aerobic oxidation of organic matter. The BOD test is a **bioassay** (BY oh ASS ay) test, meaning it provides a measure of some biological matter not by measuring the amount of matter directly but by measuring the effect of that matter on biological organisms.

The BOD test involves collection of two identical water samples from a test site. The dissolved oxygen is measured in one sample immediately. The other sample is placed in a lightproof, temperature-controlled environment for five days at 20°C. The incubation allows bacteria in the water to consume organic matter and, in the process, utilize dissolved oxygen. At the end of five days, the dissolved oxygen of the incubated sample is measured. Computing the difference between the two DO values yields the BOD of the test site.

This value indicates the amount of dissolved oxygen utilized by microorganisms as they metabolize the organic matter present in the sample (or by chemicals as they react with the oxygen) over a standard length of time at optimum conditions. Water with normal amounts of nutrients and aerobic bacteria usually loses less than 0.5 mg/L dissolved oxygen over the five days in these conditions. For most reliable results, run the test on at least three sets of samples.

Here is an example of how BOD is calculated. Assume that your initial water sample contained 12 mg/L of dissolved oxygen. You have allowed the incubated samples to sit undisturbed in a dark place for five days. During the incubation period, organic matter in the sample will be decomposed by bacteria, using dissolved oxygen during the oxidation process. The metabolism of microorganisms that steadily multiply as they feed on the organic matter will also consume oxygen.

Suppose that at the end of five days, you find that 10 mg/L of dissolved oxygen remains. If you subtract 10 mg/L from 12 mg/L, you find that your river or stream has a BOD of 2 mg of O_2 per liter. The smaller the difference between final amount of dissolved oxygen and the initial amount, the better the water quality. A BOD value of 1 mg/L is excellent and the water is considered pure. A value of 2 mg/L is good, and a value of 3 mg/L is fair. The water is considered poor if it has a BOD of 5 mg/L or more.

If you obtain a difference in dissolved oxygen over the five-day period, you do not know whether the demand comes from organic matter or chemical matter. Some tests can determine which type of source is placing the demand; for most rivers, however, the demand comes primarily from organic matter.

How Can the BOD of a River or Stream Be Improved?

Rivers with high BOD are often found to have large amounts of oxygen-depleting materials such as organic waste or chemicals introduced by factories. Human actions can improve the BOD of a river or stream by reducing or eliminating the dumping of organic waste into waterways. For instance, industries and sewage plants can partially decompose, or oxidize, their waste before releasing it into waterways. In many cases, treating wastes with oxygen renders it more biodegradable by bacteria and other microorganisms. Other plants may artificially aerate the water by bubbling oxygen through it or by increasing the flow rate.

Reducing thermal pollution can also reduce microorganisms in the waterway, which will, in turn, reduce BOD. Lower water temperatures retard bacterial growth while increasing the amount of oxygen the water can hold.

Questions

1. Explain, in your own words, the difference between BOD and DO.
2. Explain how the biochemical oxygen demand of the water might affect the organisms living in or near the river or stream.
3. Rotting organic materials such as grass clippings and leaf litter can be a mixed blessing for waters. Decaying plant materials provide food for many aquatic organisms, but they also remove oxygen. You and your family have a small pond on your land. What should you do with the lawn waste from the surrounding land?
4. How would a heavy rain affect the BOD of a river or stream?
5. Why should a river be tested for both the BOD and the DO tests? What would the results be for each if you had a high-quality water supply?

Name

STUDENT ACTIVITY 5.3

Dissolving Oxygen in Water

Purpose

To observe the effects of different aquatic mixtures on dissolved oxygen.

Background

In this activity, you will investigate conditions under which different water solutions take up oxygen. You will learn how to best collect water samples and conduct the dissolved oxygen test. You will develop skills in interpreting oxygen data in terms of percent saturation and possible needs of aquatic organisms.

As your teacher instructs, you will select a water source and an aeration method to use for this activity. Water sources may include:

- a freshwater aquarium, ready and working, with an aerating pump
- a container of water filled with **Elodea** (ehl OH dee ah) or a similar leafy, submerged aquatic plant set in the sun or under light
- a nearby pond or stream
- tap water
- distilled water (as a control)
- distilled water mixed by pouring back and forth between containers
- distilled water with ice cubes
- other in-class models of aquatic environments, such as a polluted aquarium, one with no animals, one with many fish, or one with many aerators

The aeration method you use may include:

- stirring
- an aquarium pump
- whipping with an egg beater
- pouring water back and forth between containers
- no aeration
- a method you create yourself

When you collect your water sample, use only the dissolved

Materials

Per student

- safety goggles, lab apron, and gloves

Per group

- 1L water from nearby pond or stream
- 1L dechlorinated tap water
- wide-mouth container, at least 1-L (such as glass jar, battery jar, or milk carton with top cut off)
- dissolved-oxygen bottle with stopper, 60-mL
- dissolved-oxygen test kit
- laminated instructions for the dissolved-oxygen test kit
- paper towels
- Celsius thermometer
- aluminum foil, 30 cm (8 in.)
- stirring rod or spoon
- 200 mL deionized or distilled water, if running control
- 10 ice cubes

Per class

- freshwater aquarium with aerator pump
- Elodea or similar aquatic plant, in container of chlorine-free water
- egg beater
- forceps
- waste container, 1–2 L
- first-aid kit

Optional

- additional models of aquatic environments, such as a polluted aquarium with no animals or an aquarium with many fish
- light source

oxygen (DO) bottle from your testing kit. Do not use any other kind of bottle, and do not transfer water to the DO bottle from another container.

DO bottles are specifically designed for collecting water samples that are to be tested for dissolved oxygen. The neck of a DO bottle has a collection collar that surrounds the opening. To keep air bubbles from forming inside the DO bottle, make sure that the collection collar contains water when you insert the glass stopper. No air bubbles should be trapped inside the bottle.

Once you have determined the dissolved oxygen concentration and the temperature, you will use Figure 5-1 on page 137 to obtain the percent saturation. **Percent saturation** is the percent of milligrams of oxygen gas dissolved in one liter of water at a given temperature compared with the maximum milligrams of oxygen gas that can dissolve in one liter of water at the same temperature.

Example: Assume the average temperature is 20°C and the average dissolved oxygen is 8.0 mg/L. Using Figure 5-1, locate 20°C on the Water Temperature scale and locate 8.0 mg/L on the Dissolved Oxygen Concentration scale. Using a ruler or straightedge, draw a line from 20°C to 8.0 mg/L. The drawn line will intersect the Dissolved Oxygen Saturation line at 85%.

Safety

Review safety procedures. Be sure to wear safety goggles, lab aprons, and gloves. Carefully follow the test kit manufacturer's procedures and precautions. Determining dissolved oxygen concentrations involves use of caustic materials. Dispose of all liquid DO waste in the container labeled heavy metals. Place solid wastes such as the empty chemical packets in garbage bags. Do not dispose of any wastes down the laboratory sink. If the waste jug becomes too full, pour the watery liquid into a marked, open plastic container and place the container under the hood to evaporate the water. Place the solids that remain into the "heavy metals" container.

Procedure

PART A. Planning a Dissolved-Oxygen Test

1. Based on information and materials provided by your teacher, as a group, select a water source and aeration methods to test. Using your journal, describe your chosen water source and aeration method or methods.
2. Prepare a chart in your journal that reflects all the variables you will gather during the test, such as water temperature, dissolved oxygen content, and percent saturation.
3. As instructed, inform your teacher of the procedure you plan to do.
4. Put on your safety goggles, gloves, and lab aprons. Prepare the lab table for testing.
5. Place the water sample in a 1-L container. Make sure the container holds enough sample that you will be able to submerge a DO bottle in that water.
6. Using your selected procedure, aerate your sample water source. Quantify the length of time or the number of repetitions for your actions. Record your actions on your data table.
7. Use the thermometer to measure the temperature of your sample. Record your results in degrees Celsius (°C).

PART B. Filling the Dissolved-Oxygen Bottle

8. Label a DO bottle with your group name and your method for adding oxygen.
9. Submerge the bottle in an inverted position in the water you are sampling.
10. Rotate the bottle slowly to an erect position so the water can gently enter without agitation. Avoid any rapid mixing.
11. When the bottle and the collection collar around the neck of the bottle are completely full of water, remove the bottle from the water.
12. Insert the glass stopper into the water-filled collar and push it gently into place.
13. When this procedure is properly completed, the bottle should contain no trapped air bubbles. If air bubbles are present, repeat Steps 9–12.

PART C. Testing for Dissolved Oxygen

14. Use your dissolved-oxygen test kit to determine the dissolved oxygen of your sample, following the directions for the kit. When you have completed the procedure, record the value for dissolved oxygen in mg/L, in your data table.
15. If your teacher so instructs, repeat Steps 8–14 for one or two additional aeration types or water samples, or repeat trials of the same source and procedure. Normally, in the field you would do this procedure three times for the same water source.

Observations

DATA TABLE: Dissolved Oxygen in Various Water Sources

Water Source	Aeration Procedure	Trial	°C	mg/L	% Saturation

Chart for Converting Dissolved Oxygen Concentration at a Certain Temperature to Percent Saturation

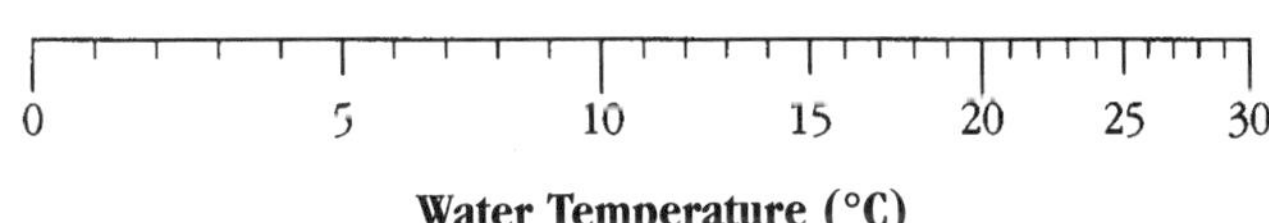

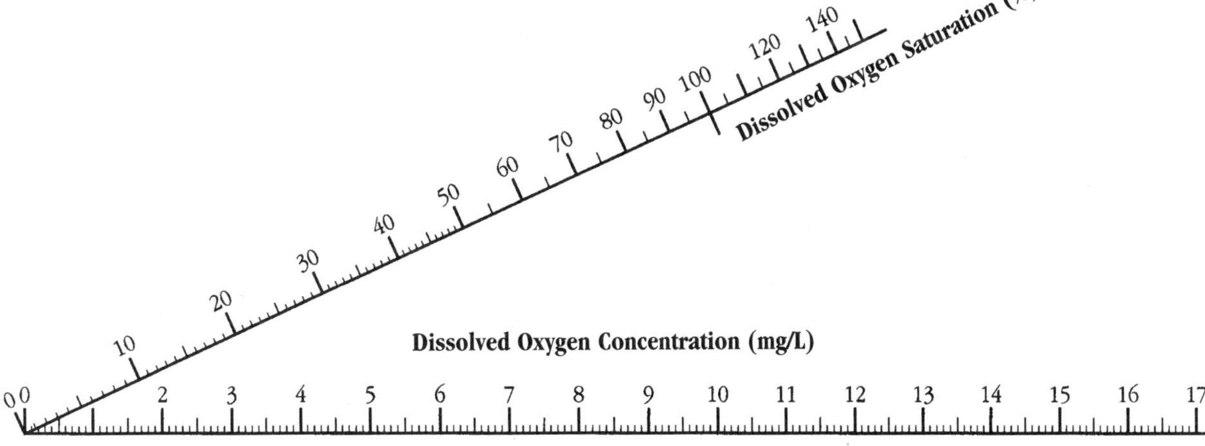

Figure 5-1: Chart for Converting DO Concentration at a Particular Temperature to Percent Saturation

Calculations

15. Using the percent saturation and temperature chart in Figure 5-1, determine the percent of oxygen dissolved in each trial of each water sample. Record these results in your data table.
16. Obtain the results obtained by other groups. Record their sample descriptions and results in your data table.

Analyses and Conclusions

1. Rank the samples from the most oxygen (in mg/L) to the least.
2. Is the percent saturation ranking the same as the mg/L ranking? Why or why not?
3. Explain why the distilled water would have no oxygen.
4. What do you think would happen to the container with the greatest amount of oxygen if you let it stand for the afternoon? Why would this happen?
5. What effect did (or would) lowering the sample temperature with ice have on the dissolved oxygen? Explain this result.
6. Discuss why more than one test is conducted on any one sample site and why you need to do multiple tests at your river or stream field site.
7. If you have data for three trials on one type of sample, account for different results.

Critical Thinking Questions

1. For each aeration technique used by your class, describe a parallel aeration force in a river or stream environment.
2. Which way of adding oxygen do you think is most practical for large amounts of water, such as a lake? Support your answer.
3. State some conditions in which using a green plant might be better than mechanical mixing of water as a way of increasing the oxygen content of an aquatic environment.
4. What types of aquatic organisms might exist in each of the sample waters?

Keeping Your Journal

1. Offer some observations of what you have learned about the effects of temperature and mixing on oxygen content of water.
2. Imagine you have a contract from a local farmer to make his small pond more healthy. You ascertain that the main problem is a lack of oxygen. Outline and illustrate what you would do to increase the oxygen supply in the water.

Name

STUDENT ACTIVITY 5.4

Measuring Dissolved Oxygen in a River or Stream

Purpose

To measure the amount of dissolved oxygen in water samples taken from a river or stream and to use this information to assess water quality.

Background

The dissolved oxygen value for your river or stream can be used as part of the Water Quality Index (WQI), which, as you have learned, gives an overall rating for river water based on nine parameters. When you collect your water samples for this test, use only the dissolved oxygen (DO) bottle from your testing kit, and do not transfer water to the DO bottle from another bottle or container. Use the procedures for collecting DO water samples presented earlier in this lesson. To get accurate dissolved oxygen readings, you must complete this activity at the field site.

Your teacher will tell you how many samples to test. Doing a second test that provides similar results demonstrates that you are doing the procedure properly. Always compare your test results to other groups.

Safety and Waste Disposal

Follow all laboratory and field safety procedures presented in Lesson 1. Be sure to wear safety goggles, gloves, and aprons when testing water that may contain contaminants. Pour all waste into a waste container.

Materials

Per student

- safety goggles, lab apron, and gloves
- life jacket and tow line (for deeper-water sites)
- rubber waders or boots (for shallower-water sites)
- Student Activity 5.3: Dissolving Oxygen in Water
- journal

Per group

- Celsius thermometer, alcohol-filled with metal jacket, or electronic
- dissolved-oxygen test kit
- dissolved-oxygen bottle with stopper
- laminated instructions for the dissolved-oxygen test kit
- water-sampling pole, with clamp
- aluminum foil, 30 cm (8 in.)
- labeling tape or marker
- plastic garbage bag (for trash)
- straightedge or ruler

Per class

- topographic map or other materials for determining location
- deionized or distilled water, about 2 L
- cooler with ice, for transporting DO bottles
- waste container, 1–2 L
- lightproof incubator set at 20°C, or lab drawer with Celsius thermometer

Procedure

PART A. Testing for Dissolved Oxygen

1. In your journal, record time, date, and location. (Use the location identification technique your teacher indicates.) Record current weather and water conditions, and note weather conditions for previous days to the extent you are aware of them.
2. Put on safety goggles, gloves, and other safety equipment for your site.
3. Measure and record the water temperature.
4. Rinse a DO bottle with river water.
5. Obtain a sample of river water in the DO bottle. Make sure no bubbles are inside. If bubbles are present, empty the bottle into the river a few meters downstream, and collect another sample, again making sure no air bubbles are inside.
6. Immediately stopper the bottle. Firmly holding the stopper in the bottle, invert the bottle and pour off excess water around the collar.
7. Using a dissolved-oxygen test kit, follow the directions provided with the kit to determine the amount of dissolved oxygen. Record in your own Data Table.
8. Empty the contents of the bottle into the waste container.
9. Rinse the DO bottle several times with deionized water or distilled water, placing the rinse water in the waste container. Rinse the DO bottle several more times with river water.
10. Repeat Steps 3, and 5 through 9 at least one more time. Record the results as Sample 2 for your group

PART B. Obtaining Water Samples for BOD Testing

11. Repeat Steps 4–6 for one DO bottle. If instructed by your teacher, repeat this for 2 or 3 bottles.
12. Completely cover the bottle with aluminum foil so that light cannot enter. Label the bottle with your group name or number and the location.
13. Place the bottle in a cooler for transport to the classroom.
14. If you have time, clean up any litter left by previous visitors to the site.

PART C. In the Laboratory

15. When you return to the classroom, set the bottle in a lightproof, temperature-controlled incubator set at 20°C or in a lightproof laboratory drawer. Record the incubator temperature on your Data Table. The bottle should remain undisturbed in the incubator for five days. You will use this sample in Student Activity 5.5 when you test for biochemical oxygen demand.

Observations

River or Stream ____________

Location ____________

Date ____________

Time ____________

Weather Conditions ____________ Weather Previous 24 Hrs ____________

Water Conditions ____________ Air temp ____ °C Flow Rate ____ m/sec

School ____________ Investigators ____________

DATA TABLE

Group	Sample	Water Temperature (°C)	Dissolved Oxygen (mg/L)	Percent Saturation
Average				

Q-value = ________ %

Initial Incubator Temperature = ________ (°C)

Calculations

16. Obtain and record the DO and temperature value results from the other groups.
17. Calculate and record the average water temperature for the class.
18. Calculate and record the average dissolved oxygen concentration for the class.
19. Using the average values temperature and dissolved oxygen, determine the percent saturation using Figure 5-1 in Student Activity 5.3 and a straightedge or ruler. Calculate the average percent saturation. Record your results.
20. Convert the average percent saturation of dissolved oxygen into its equivalent Q-value using Figure 5-2. Record the Q-value.

Figure 5-2: Graph for Converting Percent Saturation to Q-value

Analyses and Conclusions

Q-value for Dissolved Oxygen	Quality of Water
90–100%	Excellent
70–89%	Good
50–69%	Medium
25–49%	Medium
0–24%	Very bad

1. Using your Q-value for dissolved oxygen and the information in the preceding chart, rate your river or stream water from excellent to very bad.
2. Examine the results you obtained from all the groups. Were any of the reported values for DO very different from the others? How did this affect the resulting Q value? What could have caused this difference?
3. Describe your field site. Is the stream or river swift-flowing or sluggish? Do ditches or drains empty matter into your waterway? Are farms, homes, or industries nearby? Are there trees and plants that may drop leaves or other debris into the water? Is the water clear or turbid? What organisms do you see? Are there signs of decomposing matter?
4. What elements of the site you just described do you think remove oxygen from the water?

Critical Thinking Questions

1. Explain how the amount of DO in the water at your field site might affect the organisms living in the waterway.
2. Predict how the concentration of DO in the water at your field site might vary at different times of the year. Explain your reasoning.
3. Explain how temperature and turbidity might affect the DO level of a river or stream.
4. What might be done to improve the level of DO in the water at your field site?
5. If you wanted to go trout fishing, what kind of river or stream would be likely to support a population of brook trout?

Keeping Your Journal

1. Write your impressions or feeling about your field site. Do you think it is ugly or pleasing? How might it be improved? What could you do to improve the site?
2. Write a few descriptive phrases about the river or stream that you might later develop into a poem or story about your site. Sketch any unusual or interesting features of the site.

Name

STUDENT ACTIVITY 5.5

Measuring Biochemical Oxygen Demand in River or Stream Water

Purpose

To measure the level of biochemical oxygen demand in water samples taken from a river or stream and to use this information to assess water quality.

Background

In this activity you will measure the biochemical oxygen demand in the water sample you collected at the field site in Student Activity 5.4. You will obtain a final dissolved oxygen value by testing the amount of dissolved oxygen in your sample, which has been kept in darkness at 20°C for five days. To find the BOD level, you will subtract this final dissolved oxygen value from the initial dissolved oxygen value obtained at your river or stream during Student Activity 5.4.

Safety and Waste Disposal

Follow all laboratory safety procedures presented in Lesson 1. Be sure to wear safety goggles, gloves, and apron. Dispose of all liquid waste in the waste container provided by your teacher.

Procedure

1. From Student Activity 5.4, record the sampling date, location, time, air temperature, weather conditions, initial incubator temperature, and any other identification information. Also record the average initial DO level obtained.
2. Record the final incubator temperature. If the incubator did not remain at 20°C, your results will not be valid.
3. Put on safety goggles, apron, and gloves.
4. Take the DO bottle from the incubator and unwrap it. Unstop the bottle. Use a dissolved-oxygen test kit to determine the amount of DO in the sample. Record the amount of DO (mg/L) in your data table as final DO.
5. Empty the bottle contents into the waste container provided by your teacher. Rinse the bottle 2 or 3 times with tap water and dispose of the water in the waste container.

Materials

Per student

- safety goggles, lab apron, and gloves
- calculator (optional)

Per group

- water-filled DO bottle from Student Activity 5.4, incubated for five days
- dissolved-oxygen test kit
- laminated instructions for the dissolved-oxygen test kit
- waste container, 1–2 L

Per class

- lightproof incubator set at 20°C, or lab drawer with Celsius thermometer

Observations

River or Stream ______________ Location ______________

Date ______________ Time ______________

Weather Conditions ______________ Weather Previous 24 Hrs ______________

Water Conditions ______________ Air Temp ____ °C Flow Rate ____ m/sec

School ______________ Investigators ______________

DATA TABLE

Incubator Temperature (Day 1) ______ (°C)

Incubator Temperature (Day 5) ______ (°C)

Group	**Sample**	**Initial DO (Day 1) (mg/L)**	**Final DO (Day 5) (mg/L)**	**BOD (Day 1–Day 5) (mg/L)**
Average				

Q-value = ________ %

Calculations

6. Obtain and record in your own data table the initial and final dissolved oxygen results from the other groups.
7. Calculate the BOD level for each sample by subtracting the DO concentration (mg/L) obtained on Day 5 from the DO concentration (mg/L) obtained on Day 1. Record your results in your data table.
8. Using results from all groups, calculate the average BOD.
9. Convert the average BOD value into its equivalent Q-value using Figure 5-3. Record the Q-value.

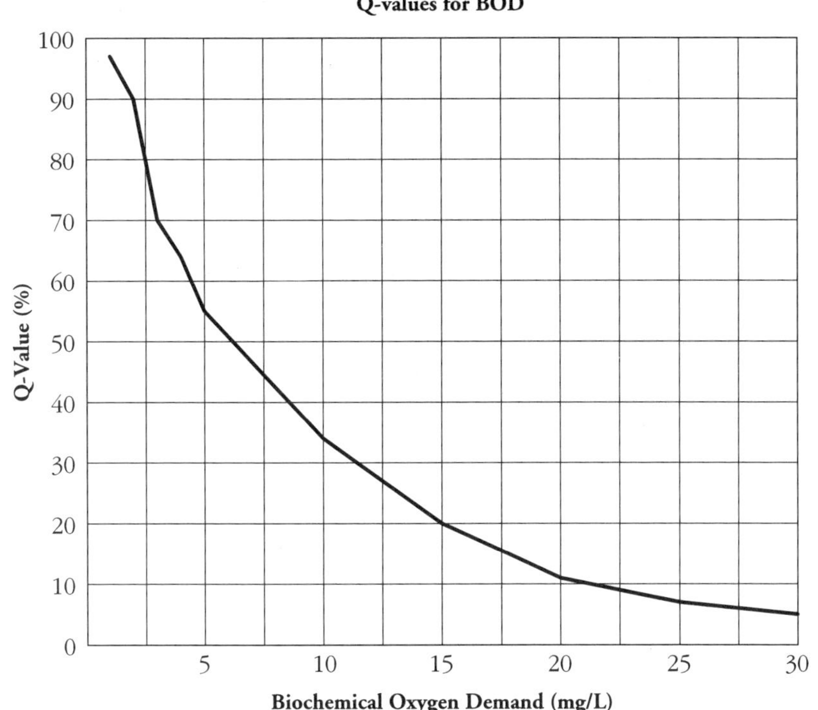

Figure 5-3: Graph for Converting BOD to Q-value

Analyses and Conclusions

Q-value for BOD	Quality of Water
90–100%	Excellent
70–89%	Good
50–69%	Medium
25–49%	Bad
0–24%	Very bad

1. Using your Q-value for biochemical oxygen demand and the preceding table, rate your river or stream water from excellent to very bad.
2. Review the description of your field site from Student Activity 5.4 (Analyses and Conclusions, Question 3). What factors do you think contributed to the biochemical oxygen demand of the water at your field site and why?
3. What might be done to improve the BOD level of the water at your field site?

Critical Thinking Questions

1. Explain how the BOD of the water at your site might affect aquatic organisms living there.
2. Predict how the biochemical oxygen demand of the water at your field site might vary at different times of the year. Explain your reasoning.
3. Explain how temperature and turbidity might affect the BOD at your site.

Keeping Your Journal

1. How did the BOD of your river or stream affect your impressions of the site?
2. Describe how this experience has affected your overall feelings about placing matter in a ditch, storm drain, or the river or stream itself.

TEACHER NOTES

LESSON 6

Fecal Coliform Bacteria in Rivers and Streams

Focus

In this lesson, students receive an introduction to the microbiological techniques to successfully isolate, culture, and enumerate bacteria, with an emphasis on bacteria of fecal origin. Students will be able to evaluate the quality of a water sample based on either fecal or total coliform content.

Learner Outcomes

Students will:

1. Learn what fecal and total coliform and other pathogenic organisms are.
2. Learn the dilution, culture, and enumeration techniques necessary for assessing fecal coliform levels in a river or stream sample.
3. Be able to describe sources of coliform bacteria.
4. Determine the amount of fecal coliform in a water source and determine the water quality of that source.

Time

Five (or up to nine) 40–50 minute class periods and a field trip to a river or stream

DAY 1:	Student Information 6.1: What Are Fecal Coliform Bacteria?
DAY 2:	Student Information 6.2: Techniques for Fecal Coliform Testing
DAY 3:	Student Activity 6.3: Measuring Fecal Coliform in a River or Stream, Part A
(OPTIONAL)	Student Activity 6.3: Measuring Fecal Coliform in a River or Stream, Part B (Analyzing Practice Samples)
DAY 4:	(Optional) Student Activity 6.3: Measuring Fecal Coliform in a River or Stream, Part B (the day after starting Part B)
FIELD TRIP:	Student Activity 6.3: Measuring Fecal Coliform in a River or Stream, Part C Student Activity 6.3: Measuring Fecal Coliform in a River or Stream, Part D (must be the same day as Part C, but in the classroom)
DAY 5:	Student Activity 6.3: Measuring Fecal Coliform in a River or Stream, Part E (must be the day after the field trip)
DAYS 6–9:	Student Assessment 6.4: Swamp Water (optional)

Safety and Waste Disposal

Follow all laboratory and field safety procedures presented in Lesson 1. Be sure students wear safety goggles, aprons, and gloves in the laboratory. Emphasize the importance of sterile technique, both to prevent contamination of samples and to avoid exposure to potentially hazardous microorganisms. Have students wash hands before and after handling materials. Washing the entire surface of lab spaces and nonsterile equipment with a 70% alcohol solution will lessen contamination during the activity. The students should never open a Petri dish that contains bacteria.

At the end of each fecal-coliform laboratory activity, collect all Petri dishes and absorbent pads, because live bacteria may still be present. Open the dishes, and place them and the absorbent pads in either 70% ethyl alcohol or, as a cheaper method, a strong bleach solution. After several days, remove the pads from the solution and incinerate them. You may dry the plastic Petri dishes and use them in such projects as observing benthic macroinvertebrates but not for bacteria cultures.

Advance Preparation

Prepare to supply students with Student Information, Activity, and Assessment sheets 6.1 through 6.4. Gather all necessary equipment and materials.

All equipment used in the fecal coliform analysis must be sterilized. An autoclave, pressure cooker, or pot of boiling, distilled water will work for sterilization. Treat all equipment for 20 minutes. After the equipment has been sterilized, it should not be handled except with gloved hands or sterilized tongs. Store the sterilized equipment in sterile, sealed, plastic bags until needed. Allow at least one hour of preparation time for this process. For best results and most efficient use of class time, do all the sterilization yourself, except have students do lab surfaces and their own hands.

At least 2 to 4 hours before students begin fecal coliform testing, adjust the water bath (or other constant-temperature environment) until it reaches a constant temperature of 44.5°C (± 0.2°C). Once the bath has reached this temperature, it can be left on for several hours until needed. Place the hot-water bath in a location least subject to temperature changes; constant temperature is critical to culturing the coliform.

If you want students to test samples before doing their own collection at the field site, provide river or stream water samples on those days. If possible, collect the sample within five hours of class usage. Also obtain a sample contaminated source of fecal coliform from the local sewage treatment plant. If possible, have them provide a count on the sample with which students can compare their own data. Before you do the field trip, review information in the unit introduction on field-trip management.

Materials

Student Activity 6.3: Measuring Fecal Coliform in a River or Stream

Per student

safety goggles, lab apron, and gloves
life jacket and tow line (for deeper-water sites)
rubber boots or waders (for shallower-water sites)
Student Information 6.2: Techniques for Fecal Coliform Testing
journal

Per group
isopropyl alcohol solution (70%)
alcohol or Bunsen burner
200-mL bottle, with stopper
3 150-mL bottles
forceps or tongs
4 ampoules fecal-coliform culture medium (about 15–20 mL total)
4 sterile filters
sterile microfiltration (MF) kit
3 25-mL pipettes
4 100-mL pipettes
hand pump for pipettes
500 mL sterile, distilled water or phosphate buffer
4 Petri dishes equipped with absorbent pads
Per class
bleach solution in a container
topographic map or other materials for determining location
cooler and ice, or refrigerator, for keeping samples cold
method to sterilize equipment such as an autoclave or pressure cooker and hot plate
hand vacuum pump that fits the MF kit
waterproof tape (such as duct tape)
sterile resealable plastic bags
waterproof marker or labeling tape
weights to submerge cultures in water bath
disinfectant soap
constant-temperature environment, such as a water bath
Optional
water-sampling pole, with clamp
microscope or hand lens
Optional, per group, if doing practice testing
100 mL distilled or boiled water, at 4°C
100 mL contaminated water, at 4°C
100 mL river or stream water, at 4°C
3 150-mL bottles, sterilized
3 100-mL pipettes
3 ampoules fecal-coliform culture medium (about 15–20 mL total)
3 sterile filters
3 Petri dishes equipped with absorbent pads

Student Assessment 6.4: Swamp Water
Per student
safety goggles, lab apron, and gloves
Student Information 6.2: Techniques for Fecal Coliform Testing
Student Activity 6.3: Measuring Fecal Coliform in a River or Stream

Per group
isopropyl alcohol solution (70%)
alcohol or Bunsen burner
200 mL "swamp water"
200-mL bottle, with stopper
2 150-mL bottles
100-mL graduated cylinder
forceps or tongs
4 ampoules fecal-coliform culture medium (about 15–20 mL total)
4 sterile filters
sterile microfiltration (MF) kit with vacuum device
3 25-mL pipettes
3 100-mL pipettes
hand pump for pipettes
500 mL sterile, distilled water or phosphate buffer
4 Petri dishes equipped with absorbent pads
Per class
bleach solution in a container
cooler and ice, or refrigerator, for keeping samples cold
method to sterilize equipment such as an autoclave or pressure cooker and hot plate
hand vacuum pump that fits the MF kit
waterproof tape (such as duct tape)
sterile resealable plastic bags
waterproof marker or labeling tape
weights to submerge cultures in water bath
disinfectant soap
constant-temperature environment, such as a water bath
Optional
microscope or hand lens

Vocabulary

colony
control plate
culture
dilution
disinfectant
enterobacteria
fecal coliform
incubation
indicator organism
medium
membrane filtration (MF)
pathogenic
phosphate buffer
Presence/Absence (P/A) test
sterile
sterilization
too numerous to count (TNC)

Background for the Teacher

Fecal coliform bacteria are universally used as indicator organisms in the determination of microbiological water quality. Among the numerous methods for determining the presence of fecal coliforms in water, the microfiltration (MF) method is most applicable for school use. Disposable MF kits are less expensive than autoclavable MF kits.

In microfiltration, a water sample is vacuumed through a latex filter, leaving the bacteria on the surface of the filter. The filter is placed in a Petri dish containing an absorbent pad saturated with a culture medium. The culture medium contains a blue stain that is specific for fecal coliforms. The culture dishes are incubated in a hot-water bath for 24 hours at 44.5°C. If you make your own hot-water bath, it must maintain water temperature to within 0.2°C.

Though having each group use its own hand vacuum pump is convenient, three groups can take turns using a single pump. This reduces the amount of special equipment needed to run this test. To assure accurate results, use only culture medium that has not expired.

Both *Rivers Biology* and *Rivers Chemistry* include fecal coliform testing. This book involves more biology knowledge and laboratory skills, however, so the *Rivers Biology* class may want to do the test and provide the results to any *River Chemistry* classes. If necessary, you can have fecal coliform analysis performed by a water-testing laboratory, public health department, or hospital. Having the students do the procedure, however, teaches them sterilization techniques and laboratory procedures that can help improve testing skills and provide good and useful data.

Introducing the Lesson

1. Have students read, discuss, and answer questions for Student Information 6.1: What Are Fecal Coliform Bacteria? (Answers for student sheets are in Appendix B.) (Some of this handout was also prepared for *Rivers Chemistry*.)
2. Have students read, discuss, and answer questions for Student Information 6.2: Techniques for Fecal Coliform Testing. (Some of this handout was prepared for *Rivers Chemistry*.) Display and demonstrate collection and sterilization equipment if desired.

Developing the Lesson

1. Set the constant-temperature environment at 44.5°C (±0.25°C), to be ready at class time.
2. Have students carry out Student Activity 6.3: Measuring Fecal Coliform in a River or Stream, Part A, in the classroom. As students carry out the activities in Lesson 6, observe and evaluate their individual performance on laboratory skills, recording your assessment on the Biological Field-Study Proficiency Checklist in Appendix A.

3. If desired, have students practice the fecal-coliform handling and measurement techniques on several different types of water samples before collecting and testing water from the field site. If so, have students carry out Part B of Student Activity 6.3 over the course of two adjoining days. You may want to provide three samples per group: germ-free water such as boiled, tap, or distilled water; water from a known contaminated site such as a sewage treatment plant; and local river or stream water. The key is to provide students with one positive sample, so they obtain results they can compare with future testing.
4. At the field site, have students carry out Part C of Student Activity 6.3. Indicate what technique students should use for determining location.
5. In the laboratory, have students carry out Part D of Student Activity 6.3.

Concluding the Lesson

1. The next day, have students carry out Part E of Student Activity 6.3 and complete the activity. Have students discuss results and answers. For student handouts on calculating the standard deviation of results, see *Rivers Mathematics*.

2. The overall WQI is not usually computed with results from only three or four water-quality tests, such as those included in *Rivers Biology*. If your class obtains results from additional water-quality tests for their site such as from a *Rivers Chemistry* class, or if you otherwise desire, you may have students compute and discuss the overall WQI for their site, using Student Activity 3.4.
3. If desired, have students read Student Assessment 6.4 as homework. Prepare the "swamp water" by placing fecal material from some herbivore, such as a rabbit, cow, or horse, into 2 L of water overnight.

Assessing the Lesson

1. Student Assessment 6.4: Swamp Water requires all or part of up to three class periods. Assign students to groups. Have students perform Part A of Student Assessment 6.4 in class. (You may want to test one sample to be sure student results are accurate.)
2. Write on the board what dilutions students should use. Indicate whether each group should do one dilution three times, do three different dilutions, or choose a dilution factor based on their own experience. Have them complete Part B of Student Assessment 6.4. As the groups work, check them for technique and safety.
3. Have students do Student Assessment 6.4 Part C and complete this activity. As the groups work, check them for technique and safety. Draw a data table on the board and have each group list their results. Here is a suggested scoring rubric, which you can use with the performance criteria included in the student sheet.

Scoring Rubric

Score	Expectations
0	Some individuals do not outline safety procedures correctly. Group does not complete MF testing properly. Does not explain the results.
1	Most members outline safety procedures properly. Group has some problem completing MF testing. Written work explains some analysis but is incomplete or offers inaccurate conclusions; problems in sentence structure, spelling, and grammar.
2	All members outline safety procedures properly. Group has few problems completing MF testing. Explains the analysis; outlines conclusions simply; some problems in sentence structure, spelling, and grammar.
3	All members outline safety procedures properly. Group has no problems completing MF testing. Explains the analysis; outlines and elaborates on conclusions; few problems in sentence structure, spelling, or grammar.

4	All students provide properly outlined safety procedures. The group has no problems completing the MF testing. Explains the analysis, elaborating on conclusions in detail; no problems in sentence structure, spelling, or grammar.

4. Review and assess student performance on the laboratory skills demonstrated during this lesson, as recorded on Biological Field-Study Proficiency Checklist, in Appendix A.

Extending the Lesson

The first two options listed here provide wonderful ways for students to see the testing process applied in real life.

1. Invite a sewage-treatment professional to discuss sewage treatment and fecal coliform bacteria.
2. Visit a sewage or a water treatment plant to observe fecal coliform testing.
3. Have students research what happens to sewage after it leaves their homes.
4. If fecal coliform counts for the water at the river or stream site are high, do sampling at an upstream site to try to determine the source of the contamination.
5. Purchased meat, especially poultry, may contain coliform bacteria. Have students wash chicken collected from grocery stores and meat markets in sterile water, then test the water for coliform bacteria. Students may also evaluate and discuss food processing, handling, and cooking techniques.
6. Have students do research papers on various waterborne infectious diseases. Sample topics include: diarrhea, cholera, giardia, and cryptosporidium.

What Are Fecal Coliform Bacteria?

One of the primary human concerns with water quality focuses on the presence or absence of **pathogenic** (path uh JEHN ihk), or disease-causing, organisms, particularly bacteria. Many species of bacteria live in unprotected water supplies that may look clean. Many of them are harmless, but others are not.

Certain rod-shaped, waterborne bacteria that propagate in the digestive systems of humans and other warm-blooded animals, known as **fecal coliforms** (FEE kuhl KOH luh forms), often occur in waterways. If present in drinking-water sources, they and associated bacteria can cause a variety of disorders.

Though many fecal coliforms do not cause disease, some readily establish themselves in the lower gut of vertebrates, especially humans. Once established, they produce waste products toxic to the host organism. These toxins may result stomachaches, chills, severe fever, and diarrhea. In recent years, outbreaks of cryptosporosis took place in some U.S. water systems. In many developing countries, cholera, an intestinal infection caused by a waterborne bacterium, is a hidden killer. Diarrhea is the number-one killer of infants worldwide; it often results from mixing powdered milk with water contaminated with a disease organism.

Though fecal coliforms are present in many vertebrates, little evidence suggests they harm nonhuman vertebrates. So their presence has health significance only to humans.

What Are Fecal Coliform Bacteria?

The many type of gut-dwelling bacteria known as **coliform** bacteria belong to the family Enterobacteriaceae (ehn TEHR oh bak TEER ey ah see ay). **Enterobacteria** (enh TEHR oh bak TEER ey ah) are described by the World Health Organization as rod-shaped, gram-negative (meaning it does not respond to a test known as Gram's bacterial stain), and nonspore forming.

Determining the presence of fecal coliform requires differentiating them from other coliform. Temperature holds the key to this. All Enterobacteria ferment lactose, a sugar, if held at 35°C–37°C for 24 to 48 hours. At 44.5°C (± 0.25°C), however, fecal coliform still ferment lactose, while other coliforms die or become inactive. The waste products of lactose fermentation turn fecal coliform bacteria blue, making them identifiable. So scientists utilize these temperature parameters as part of a test to determine fecal contamination of water.

Why Test Water for Fecal Coliform?

Certain bacterial groups are excellent environmental-quality indicators. The presence of such **indicator organisms** tends to indicate other events or organisms. Investigators discovered that, when outbreaks of typhoid, cholera, dysentery, hepatitis, and other waterborne pathogenic diseases occurred, high fecal coliform counts were often present. Though fecal coliform bacteria normally do not cause disease, they often coexist with pathogenic organisms. This identification of fecal coliforms as indicator organisms was a landmark in maintaining clean drinking water.

Pathogenic bacteria usually occur in small numbers and cannot exist outside a host body for very long, so they are difficult to detect. Fecal coliform, however, can exist in greater numbers and for longer periods outside the intestinal tract. This makes them easier to detect and monitor. If testing of a water source reveals the presence of fecal coliform, one can assume human or other vertebrate feces have entered the aquatic system and that pathogenic bacteria may also be present.

If more specific testing of a contaminated water sample from a river or stream reveals the presence of **Escherechia coli** (esh eh ree KAY ah KOH ly), the specific fecal coliform bacterium most commonly found in human and animal waste, then such waste is the specific, more likely source of the contamination. Not only do such wastes represent a potential health threat when they enter a water source, they also add organic nutrients, which excessively enrich aquatic ecosystems. The presence of excessive nutrients frequently results in rapid eutrophication and oxygen depletion. Such conditions may result in fish kills and the loss of other aquatic, aerobic life forms.

How Do Fecal Coliforms Enter the Water?

Broken sewer lines, faulty septic systems, and direct wastewater discharges may introduce untreated animal and human fecal matter into a waterway. Farm animals drinking directly from a stream will add their waste to the water supply.

Climatic conditions have a direct bearing on coliform counts. Excessive rainfall usually results in lower counts due to dilution by rain water. Heavy rains with runoff sometimes raise fecal coliform counts by washing in animal wastes from the land. Heavy rain or runoff into overtaxed combined storm-sewer drainage systems may cause raw sewage to flow directly into a river or stream. Extended drought conditions may also produce higher counts, because little dilution is taking place. Because 44.5°C is the optimum temperature for fecal coliforms, counts tend to be lower in seasons when water is cooler.

Standards for Fecal Coliform Levels in Water

The National Primary Drinking Water Regulations established standards for maximum concentration of specific contaminants allowed in drinking water. The standards apply to all potable water systems in the United States. Drinking water should contain zero fecal coliforms. The values listed in Table 6-1 are the current, generally accepted standards listed by the U.S. Environmental Protection Agency.

Public water systems and sewage-treatment systems test their output before release, treating it if necessary. Recreation areas also periodically test for fecal coliform levels. If the levels are too high, the site is closed for recreation.

TABLE 6-1

Water-Quality Standards

Category	**Fecal Coliforms per 100 mL of Water**
Drinking water	0 colonies
Swimming	200 colonies
Boating	1000 colonies
Treated sewage	200 colonies

Questions

1. Why is water tested for fecal coliforms rather than for pathogenic bacteria?
2. List some diseases associated with waterborne pathogenic bacteria.
3. How do fecal coliforms get into water?
4. List the fecal coliform standards for drinking water and for swimming water.
5. What characteristics of fecal coliform bacteria make them useful as indicator organisms?
6. What level of fecal coliforms do you expect at your site? Why?

Techniques for Fecal Coliform Testing

Water-quality professionals use several types of tests to determine the fecal-coliform status of a water sample. One test evaluates the presence of absence of fecal coliforms. Another quantifies the level of fecal coliforms present. In this lesson, you will learn about both kinds of tests. Any testing involves careful preparation and handling of equipment and materials. Such precautions help ensure valid results and the safety of individuals performing the tests.

Because bacteria are small and numerous, counting how many individuals are in a water sample is very difficult. To assess the level of fecal coliforms in water samples, scientists **culture** the bacteria, or propagate them in a specially prepared environment that enhances growth. The bacteria are grown on or in a medium. A **medium** is a nutrient-balanced substance on which microorganisms are grown. Each bacterium divides and grows into a **colony,** a group of bacteria. Unlike bacteria, bacterial colonies are visible to the naked eye and can be counted relatively easily.

Checking for the Presence of Fecal Coliform

Water from even the cleanest river or stream may contain at least some level of fecal coliform bacteria, because the waste products of animals in the watershed has entered the water. At low levels, it may not harm aquatic organisms. For human health, however, drinking water should contain absolutely no fecal coliforms. In testing water that must be absolutely clean, investigators can use a simple, quick screening method. This **Presence/Absence (P/A) test** indicates whether zero coliform organisms are present. This test does not indicate numbers of bacteria present (so it is useful only when the sole acceptable result is zero coliforms, such as in testing drinking water.) A water sample is mixed with testing material, then incubated for 24 hours at 35°C (±0.5°C), the optimal temperature for coliform growth. If the sample changes color, coliform bacteria are present. Enumerating fecal coliforms or other coliform bacteria requires further testing.

Counting Fecal Coliform

Though researchers have devised numerous methods of determining the abundance of fecal coliforms in water, the **membrane filtration (MF)** method is most applicable for use in secondary schools. Government standards require that microfiltration for fecal coliform testing be conducted with 100 mL of water. So, when you do this test, you will vacuum a 100-mL water sample through a latex filter, so any bacteria remain on the surface of the filter. You will place the filter in a Petri dish containing an absorbent pad saturated with a culture medium that has a stain specific for fecal coliforms.

In order to provide optimal conditions for growth and development of your testing material, you will keep it in an environment of controlled temperature, a process referred to as **incubation.** You will place the Petri dish in a hot-water incubator for 24 hours at 44.5°C. This temperature is higher than that for the P/A test; at the lower temperature, the P/A test can reveal the presence or absence of bacteria from the entire coliform group, while, at the higher temperature, all nonfecal coliforms will die.

After incubation, you will examine the surface of the filter for stained bacterial colonies. Each colony represents one bacterium from the original water sample that has multiplied to form a visible mass. By comparing the number of colonies in the sample to the water-quality standards in Table 6-1 (in Student Information 6-1), you can determine whether the water meets the standard for specific intended usage.

The number of colonies is also used to calculate the water-quality value for fecal coliforms, which contributes to the overall Water-Quality Index (WQI). The fecal coliform value is weighted more heavily than other variables in the overall WQI because it is such an important indicator of water quality for human use.

The Importance of Sterilization Techniques

Conducting any type of microbiological analysis requires maintenance of 100-percent sterility during isolation and culturing. A **sterile** surface is one that lacks any form of life, even microscopic. Any contaminated equipment would introduce "alien" bacteria, ultimately resulting in invalid data. So you must sterilize all equipment and surfaces that may affect the sample. **Sterilization** (sterhr uh luh ZAY shuhn) is a process of creating an environment free from unwanted microorganisms.

Glass, metal, and temperature-resistant plastic items can be sterilized by autoclaving them at 120°C for 15 to 20 minutes. If an autoclave is not available, a home pressure-cooker works if a pressure of 15 psi is maintained for 20 minutes. If neither autoclave nor pressure cooker is available, place items in a high-temperature oven or boiling water for 20 minutes. Metallic and glass items, such as forceps, can be flamed or boiled. Before lighting an alcohol lamp or Bunsen burner for flaming, make sure no flammable alcohol vapors are in the area.

Wash the entire surface of lab spaces and nonsterile equipment with a 70% alcohol solution to minimize contamination during the activity. (Be careful; alcohol is flammable.) Wash hands with 70% alcohol and a **disinfectant** soap (meaning one that inhibits or destroys microorganisms) before and after handling materials.

Sampling Techniques Promote Accurate Results

When you will be conducting a fecal coliform test on water from a nearby river or stream, the samples should be collected in presterilized containers, using care not to contaminate the sample with foreign bacteria from hands or other nonaquatic sources. Large dissolved-oxygen bottles with ground glass stoppers make excellent collecting bottles. Once autoclaved, the bottles remain sterile until opened.

To take water samples, submerge the collection bottle directly in the current of the waterway while facing upstream. Make sure the opening of the bottle is well below the surface. Avoid stirring up the water or the stream bed, which may introduce sediment or other organic matter into the collection container. Always use proper safety procedures when obtaining the water, such as wearing life jackets and a tow line in deeper-water sites. In heavily contaminated or suspect water, wear waders and rubber gloves.

Because cooling slows metabolism and reproduction of bacteria, refrigerate water samples collected in the field, or place them on ice, until you conduct testing. In all cases, you must conduct testing within five hours of collection.

As discussed in Student Information 6.1: What Are Fecal Coliform Bacteria?, climatic and weather conditions influence coliform counts. Whenever collecting river or stream samples, note weather and water conditions. Visually estimate contamination of the water; water that looks contaminated probably is, so compare your observations with actual data.

Dilution Techniques Enable Useful Results

As mentioned earlier, government standards require membrane filtration (MF) be conducted with 100 mL of water. In highly contaminated water, however, that volume often produces so many colonies that they are difficult to count, a condition known as **too numerous to count (TNC).** A count of 20–80 colonies per membrane is ideal; obtaining such a count often requires diluting the original sample before incubating. **Dilution** (di LOO shuhn) is a process that reduces the concentration of matter per unit volume.

The American Water Works Association (AWWA) and the Water Pollution Control Federation (WPCF) have developed recommended sample volumes and dilutions for various types of sample sources. To increase the likelihood of accurate results, dilute samples as described on Table 6-2, based on their recommendations. A particularly contaminated sample may require experimentation to determine the most appropriate dilution. For most accurate results, most procedures involve running three samples through the MF test.

TABLE 6-2

Recommended Sample Volumes

Source	Volume of Source Water in 100 mL of Testing Solution	Ratio of Source Water to Total Volume
Drinking water	100 mL	1:1
Natural bathing water	50 mL	1:2
River water	10 mL	1:10
Storm-water runoff	1.0 mL	1:100
Untreated sewage	0.1 mL	1:1000

To make a sample for dilution:

- To make 100 mL of a 0.1 mL (1:10) dilution, into a sterile container place 10 mL of water sample and 90 mL of sterile water or phosphate buffer. A phosphate buffer is a phosphate compound dissolved in water, similar to bacteria's natural environment. This buffer is preferable to sterile water, which causes bacteria to swell and die, reducing count accuracy.
- To make 100 mL of a 0.01 mL (1:100) dilution, into a sterile container place 10 mL of the 1:10 sample along with 90 mL of sterile water or phosphate buffer.
- To make 100 mL of a 0.001 mL (1:1000) dilution, into a sterile container place 10 mL of the 1:100 sample along with 90 mL of sterile water or phosphate buffer.

These sample dilution ratios are modified in Figure 6-1. Using diluted samples means that the number of colonies counted in the sample after incubation does not reflect the number of colonies per 100 mL of water sample, the government standard for evaluation. To determine actual diluted colony counts, use this formula:

$$\frac{\text{colonies counted}}{\text{dilution factor (such as 10, 100, 1000)}} \times 100 = \text{colonies per 100 mL of actual sample}$$

Example: 20 colonies counted in a 1:10 dilution would indicate:

$$\frac{20}{10} \times 100 = 200 \text{ colonies per 100 mL of actual sample}$$

Safety and Waste-Disposal Procedures for Fecal-Coliform Analysis

Any bacteriological procedure must begin with wearing safety goggles, aprons, and gloves in the laboratory. It also requires the sterilization techniques described earlier, both to prevent contamination of samples and to help avoid exposure to potentially hazardous microorganisms. The Petri dishes will contain live bacteria, so do not open them after completing your analysis. Collect all Petri dishes for disposal by your teacher.

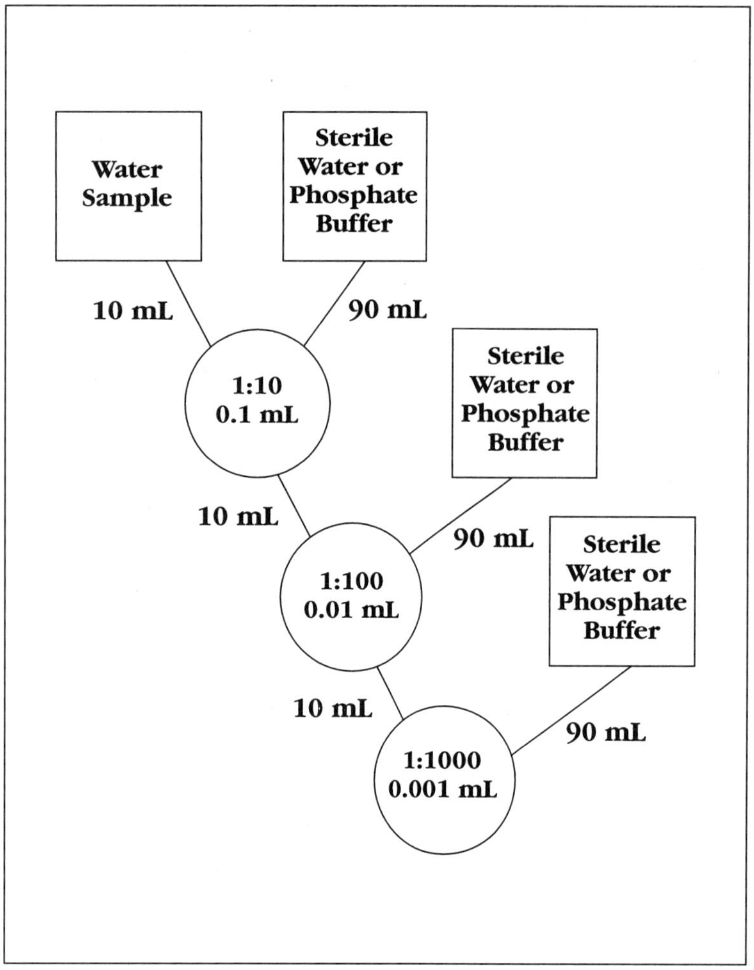

Figure 6-1, Model of Sample Dilution Ratios

Questions

1. Why should you collect samples for fecal coliform testing in sterile bottles?
2. Why must the water samples be refrigerated if testing for fecal coliform cannot be done immediately? How soon after collection must samples be tested?
3. List the procedures you would use in order to maintain sterile materials and equipment for fecal coliform testing.
4. Why should you not open a Petri dish that has been cultured for fecal coliform?
5. What do you need to consider when making a dilution to test for fecal coliform? Use some of the dilution-chart information to explain your answer.
6. What should tests be conducted on sterile or distilled water before doing an MF test?

Name

STUDENT ACTIVITY 6.3

Measuring Fecal Coliform in a River or Stream

Purpose

To collect samples of river or stream water, measure the level of fecal coliform bacteria, and assess the impact on water quality.

Background

In this activity, you will collect a sample from a river or stream and determine how many fecal coliform bacteria it contains. (Your teacher may also have you test other samples before testing your river or stream.) All fecal coliform analysis must be run within 4 to 6 hours of sample collection, and samples must

Materials

Per student

- safety goggles, lab apron, and gloves
- life jacket and tow line (for deeper-water sites)
- rubber boots or waders (for shallower-water sites)
- Student Information 6.2: Techniques for Fecal Coliform Testing
- journal

Per group

- isopropyl alcohol solution (70%)
- alcohol or Bunsen burner
- 200-mL bottle, with stopper
- 3 150-mL bottles
- forceps or tongs
- 4 ampoules fecal-coliform culture medium (about 15–20 mL total)
- 4 sterile filters
- sterile microfiltration (MF) kit
- 3 25-mL pipettes
- 4 100-mL pipettes
- hand pump for pipettes
- 500 mL sterile, distilled water or phosphate buffer
- 4 Petri dishes equipped with absorbent pads

Per class

- bleach solution in a container
- topographic map or other materials for determining location
- cooler and ice, or refrigerator, for keeping samples cold
- method to sterilize equipment such as an autoclave or pressure cooker and hot plate
- hand vacuum pump that fits the MF kit
- waterproof tape (such as duct tape)
- sterile resealable plastic bags
- waterproof marker or labeling tape
- weights to submerge cultures in water bath
- disinfectant soap
- constant-temperature environment, such as a water bath

Optional

- water-sampling pole, with clamp
- microscope or hand lens

Optional, per group, if doing practice testing

- 100 mL distilled or boiled water, at 4°C
- 100 mL contaminated water, at 4°C
- 100 mL river or stream water, at 4°C
- 3 150-mL bottles, sterilized
- 3 100-mL pipettes
- 3 ampoules fecal-coliform culture medium (about 15–20 mL total)
- 3 sterile filters
- 3 Petri dishes equipped with absorbent pads

remain refrigerated. (If absolutely necessary, samples can be kept refrigerated and the analysis run within 24 hours of collection.) In Petri dishes, you will culture a control and three prepared samples at 44.5°C (± 0.2°C) for 24 hours. (The Petri dish with the control is called a **control plate.**) After the incubation, you will count the number of colonies.

To make counting fecal coliform in your river or stream water sample easier, you will do 1:10 and 1:100 dilutions. If you analyze the same waterway several times, you will find a particular dilution that works best for your river or stream. (Aim for a dilution that yields 20 to 80 fecal coliform colonies per culture.)

Safety and Waste Disposal

Follow all field safety and laboratory procedures presented in Lesson 1 and in Student Information 6.2. Avoid direct contact with the sample. Use a hand pump rather than a mouth pipette to transfer liquids.

Follow sterilization procedures presented in Student Information 6.2 for laboratory surfaces and equipment. After completing the analysis, clean your hands well with 70% alcohol and a disinfectant soap. At no time open the sealed Petri dishes. At the end of the activity, give the unopened Petri dishes to your teacher for disposal.

Procedure

PART A. Preparing Equipment

1. Put on safety goggles, lab apron, and gloves.
2. Sterilize all laboratory surfaces.
3. Sterilize the 200-mL bottle and its stopper. Stopper the bottle and place it in a resealable plastic bag labeled with your group name or number. You will need this at the field site.
4. Sterilize a microfiltration (MF) kit, 2 150-mL bottles, 100-mL graduated cylinder, 3 25-mL pipettes, 4 100-mL pipettes, 4 Petri dishes, 3 150-mL bottles, and forceps. Place each item in a resealable plastic bag labeled with your group name or number. You will use these in the laboratory. (Sterilize additional 3 150-mL bottles and 3 100-mL graduated cylinders if testing trial samples before going to the field.)
5. Set incubator or hot-water bath at 44.5°C (±0.2°C), so it will be ready when you need it.

PART B. Analyzing Practice Samples

6. If instructed by your teacher, transfer 100 mL of distilled water to a 150-mL sterilized bottle labeled "Sample 1." Repeat for 100 mL of contaminated water sample as "Sample 2" and for 100 mL of river or stream water labeled "Sample 3."
7. Check that the incubator or hot-water bath is at the correct temperature, 44.5°C (±0.2°C).
8. Carry out Steps 23 through 29 for each sample, except label each Petri dish with the correct sample number, as well as your group name or number. Complete the day's lab with Steps 31 and 32.
9. The next day, carry out Steps 33 through 37, creating a data table of results.

PART C. Collecting a Water Sample at the Field Site

10. In your journal, record time, date, and location. (Use the location identification technique your teacher indicates.) Record current weather and water conditions, and note weather conditions for previous days, to the extent you are aware of them.
11. Describe your field site. Do ditches or drains empty into your waterway? Are farms or stockyards nearby? How does the water look and smell?
12. Put on safety goggles, gloves, and other safety equipment.
13. Clamp your sterile 200-mL sampling bottle to the water-sampling pole. Unstop the bottle, and obtain a water sample. (If your teacher directs you to, use a sampling pole to collect water away from the shore.)
14. Place the stopper back in the bottle and place in a refrigerated container. Return to the lab.

PART D. Incubating the Samples (Must Be Within Four Hours of Sample Collection, or Within 24 Hours of Field Trip if Necessary)

15. Put on safety goggles, lab apron, and gloves.
16. Sterilize all laboratory surfaces.
17. Check that the incubator or hot-water bath is at the correct temperature, 44.5°C (±0.2°C).
18. Label two sterile 150-mL containers with your group number and "1:10 Dilution" and "1:100 Dilution," respectively.
19. To make 100 mL of the 0.1 mL (1:10) dilution, use a sterile 25-mL pipette to transfer 10 mL of water sample to the sterile container labeled "1:10 Dilution." Using a separate 25-mL pipette, transfer 90 mL of sterile water or phosphate buffer to that container.
20. To make 100 mL of the 0.01 mL (1:100) dilution, use a sterile 25-mL pipette to transfer 10 mL of the 1:10 sample to the sterile container labeled "1:100 Dilution." Using a separate 25-mL pipette, transfer 90 mL of sterile water or phosphate buffer to that container.
21. Empty contents of container labeled "1:10 Dilution" down sink. Repeat Step 19, so you have 100 mL of 1:10 dilution.
22. Place the dilutions on sterile work area next to the MF kit.
23. With the sterile forceps, place a sterile filter pad in your MF kit.
24. Using a sterile 100 mL pipette, transfer 100 mL of distilled water to the MF device. (A sample of distilled water is called a **control plate.**)
25. Vacuum that 100 mL through the filter pad, following instructions supplied with your kit or by your teacher.
26. Open a sterile Petri dish equipped with a sterile absorbent pad. Add one ampoule (about 2–4 mL) of fecal coliform growth medium to the pad.
27. With sterile forceps, remove the filter pad from the MF kit and place it directly on top of the saturated pad inside the Petri dish.
28. Close the dish and seal it with waterproof tape. Using a waterproof marker, label the lid of the dish "Control Plate." Place the dish in a sterile, waterproof plastic bag. Seal the bag and label it "Control Plate" with your group name or number.
29. Place the dish upside down (to avoid condensation) in the hot-water incubator, submerging the bag with a weight. Incubate the dish for 24 hours at 44.5°C (±0.2°C).
30. Repeat Steps 23 through 29 with the undiluted river or stream water, the 1:10 dilution, and the 1:100 dilution. Label each dish accurately.
31. Wash your hands with disinfectant soap, wipe down the area with 70% alcohol, rinse your equipment with water, and place it in the area to be ready for sterilization.
32. Record any procedural questions or observational comments in your journal. Note things that did not go well or that might help when testing river or stream water.

PART E. Counting the Coliform Colonies

33. After the incubation period (24 hours) has elapsed, remove the control Petri dish from the incubator. Examine the filter surface of the Petri dish for the presence of bluish spots. Each blue spot represents one fecal coliform colony that grew from one fecal coliform bacterium. If the control plate reveals any sign of growth on the filter surface, then the equipment used was not sterile. Therefore, all other dishes will provide invalid results; proceed directly to Steps 36 and 37. Ask your teacher whether you should repeat Part C, with resterilized equipment, or whether you should use data collected by other groups.
34. If the control plate shows no colonies, remove from the incubator a Petri dish with another sample. Examine the filter surface for the presence of bluish spots. Count the number of blue spots. Other colored spots are not coliforms, so do not include them in your count. To avoid color changes, be sure to count colonies within 20 minutes of removing the dish from the incubator. The colonies should be visible, but you can use a dissecting microscope or hand lens to enlarge them. Record in your Data Table 1 the number of blue colonies in the dish. If you are not sure of

the count, have someone in another group confirm the count. When heavy contamination is present, colonies may be so close that they merge. In this case, a dish with additional dilution will provide more accurate information.

35. Repeat Step 34 for each Petri dish.
36. Give all Petri dishes and contents to your teacher to be placed in a container of disinfectant (household bleach or 70% alcohol).
37. Wash your hands with disinfectant soap, wipe down the area with 70% alcohol, rinse your equipment with water, and place it in the area to be ready for sterilization.

Observations

River or Stream ______________ Location ______________

Date ______________ Time ______________

Weather Conditions ______________ Weather Previous 24 Hrs ______________

Water Conditions ______________ Air Temp ____ °C Flow Rate ____ m/sec

School ______________ Investigators ______________

DATA TABLE

Solution Tested	Actual Fecal Coliform Count (colonies/100 mL)	Fecal Coliform Count Equivalent to Undiluted Sample (colonies/100 mL)
(A) Sterile distilled water		//////////
(B) Sample 1 (undiluted)		
(C) Sample 2 (1:10 dilution)		//////////
(D) Line (C) × 10	//////////	
(E) Sample 3 (1:100 dilution)		//////////
(F) Line (E) × 100	//////////	

Total Equivalent FC Count For All Samples (Lines B, D, and F) = ________ colonies/___ mL

Average Equivalent FC Count For All Samples = ________ colonies/100 mL

Q-Value = ________ %

Calculations

1. Using the result from Lines B, D, and F, calculate the average colonies of fecal coliform per 100 mL and record.
2. Convert the average number of fecal coliform colonies per 100 mL into its equivalent Q-value percent by using Figure 6-2. Record the result in your data table.

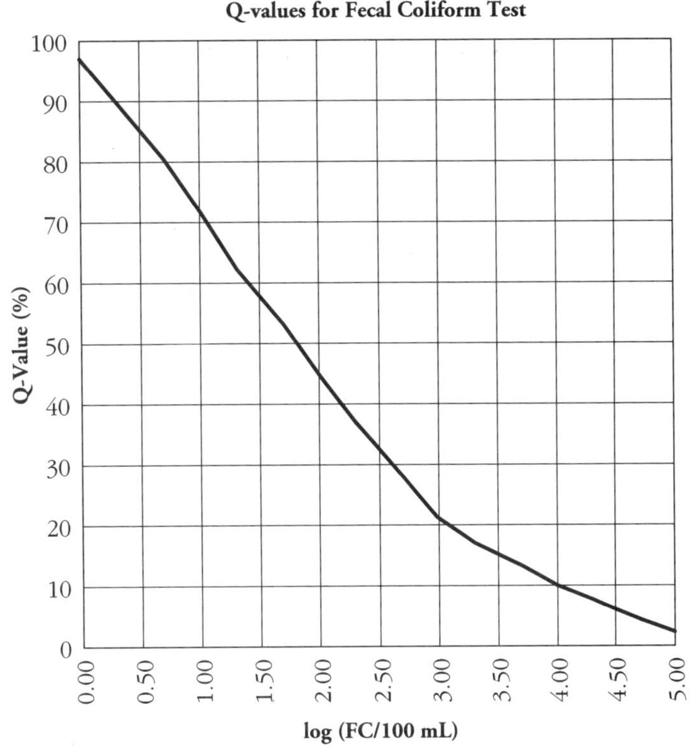

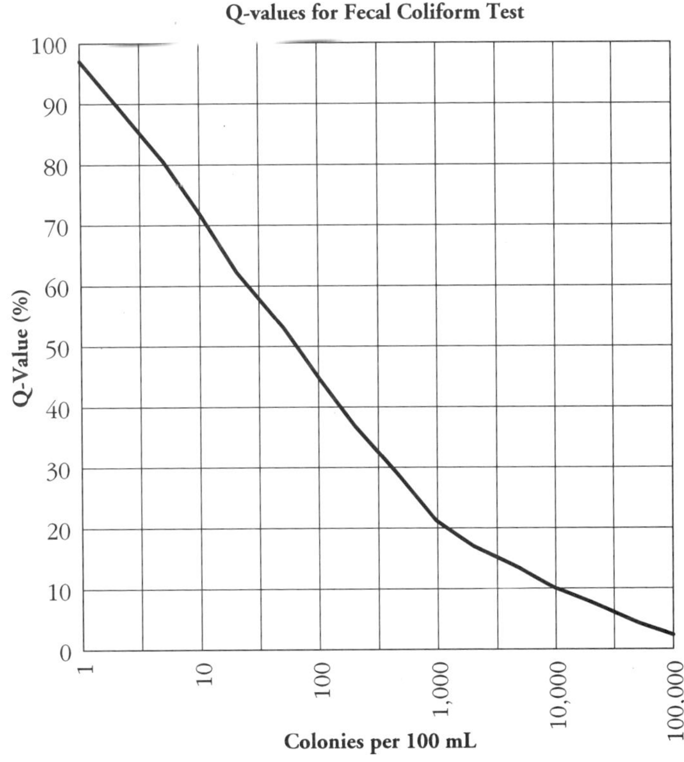

Figure 6-2: Graphs for Converting Fecal Coliform Data to Q-value

Analyses and Conclusions

Q-value for Fecal Coliform	Quality of Water
90–100%	Excellent
70–89%	Good
50–69%	Medium
25–49%	Bad
0–24%	Very bad

1. Using your Q-value for fecal coliform and the preceding chart, rate your river or stream from excellent to very bad.
2. Referring to the water quality standards in Student Information 6.1, is the water in your river or stream safe to drink? Is it safe for swimming? Is it safe for boating?
3. Describe the major natural and human-made features of your site. Begin by describing its predominant type, such as wooded, clear-cut, industrial, farmland, park, residential, or wetland. What features may be sources of fecal coliforms at your site?

Critical Thinking Questions

1. What level of fecal coliforms did you expect at your site? Was your test result better or worse than you expected? Why?
2. Explain how the fecal coliform level at your site might affect the organisms living in or near the waterway.
3. Predict how fecal coliform levels might vary at your site at different times of the year. Explain your reasoning.
4. Under what circumstances might the P/A test be preferred over the MF test to test river water? Why?
5. What would you recommend to improve the fecal coliform levels and water quality at your site? Support your opinion.
6. Students analyzed several samples of water and found the following data:

Sample Site Number	Type of Water	Fecal Coliforms per 100 mL
1 Well water	4 colonies	
2 Lake water	2800 colonies	
3 River water	1250 colonies	
4 City water	3 colonies	
5 Lake water	120 colonies	
6 River water	800 colonies	

a. Based on the preceding test results, which sites have water that would be safe to drink? Explain.
b. What might have caused the fecal coliform levels at each site?
c. What should be done with the two drinking-water supplies they sampled?
d. Water was found to have fecal coliforms present. You need to use it as a drinking-water supply. Give two or three procedures for making the water safe.

Keeping Your Journal

1. Write your impressions or feelings about your field site.
2. People will be building and living around your local river or stream site in the coming years. How will this population impact the fecal levels of that water?
3. Write a few descriptive phrases about the river or stream that you might later develop into a poem or story about your site. Sketch any unusual or interesting features of the site.

Name

STUDENT ASSESSMENT 6.4

Swamp Water

Introduction

You have been working on bacteriological studies and techniques for sampling and culturing fecal coliform bacteria in river water. Now you will show you have learned the procedure and can replicate it. In this assessment, you will be working with swamp water, a sample of water contaminated with feces from some herbivore, such as rabbit, cow, or horse. You will be expected to follow sterile procedures, use dilution techniques, carry out membrane filtration to separate the fecal coliform, and count the colonies. In following this evaluation procedure, work slowly and carefully, use your journal to check the procedures, and remember to do everything safely and use good techniques for preventing microbial contamination or infection.

Safety and Waste Disposal

Follow all field safety and laboratory procedures presented in Lesson 1 and in Student Information 6.2. Because some bacteria in a water sample may be pathogenic, avoid direct contact with the sample. Use a hand pump rather than a mouth pipette to transfer liquids.

Follow sterilization procedures presented in Student Information

Materials

Per student

- safety goggles, lab apron, and gloves
- Student Information 6.2: Techniques for Fecal Coliform Testing
- Student Activity 6.3: Measuring Fecal Coliform in a River or Stream

Per group

- isopropyl alcohol solution (70%)
- alcohol or Bunsen burner
- 200 mL "swamp water"
- 200-mL bottle, with stopper
- 2 150-mL bottles
- 100-mL graduated cylinder
- forceps or tongs
- 4 ampoules fecal-coliform culture medium (about 15–20 mL total)
- 4 sterile filters
- sterile microfiltration (MF) kit with vacuum device
- 3 25-mL pipettes
- 3 100-mL pipettes
- hand pump for pipettes
- 500 mL sterile, distilled water or phosphate buffer
- 4 Petri dishes equipped with absorbent pads

Per class

- bleach solution in a container
- cooler and ice, or refrigerator, for keeping samples cold
- method to sterilize equipment such as an autoclave or pressure cooker and hot plate
- waterproof tape (such as duct tape)
- sterile resealable plastic bags
- waterproof marker or labeling tape
- weights to submerge cultures in water bath
- disinfectant soap
- constant-temperature environment, such as a water bath

Optional

- microscope or hand lens

6.2 for disinfecting all laboratory surfaces and equipment. After the analysis is completed, clean your hands thoroughly with 70% alcohol and a disinfectant soap. At no time should you open the sealed Petri dishes. At the end of the procedure, give the unopened Petri dishes to your teacher for disposal.

Procedure

PART A. Preparing for Testing

1. List the procedures for properly collecting a sample at a river or stream.
2. List the safety procedures for conducting fecal coliform tests in the laboratory.
3. As a group, decide who will do each part of the fecal coliform test. Each member should take part in the procedure. Work together to make sure each student can perform all the procedures for which he or she is responsible, so you can all succeed. Review activity sheets and materials lists as appropriate.
4. As directed by your teacher, prepare your lab materials for the testing procedure. Use Student Activity 6.3, Part A, as a guide.

PART B. Incubating the Samples

5. Give your teacher a list of the group members and their individual responsibilities. Throughout the rest of your assessment, if one person is missing from the group, another group member must take that part. Record data in appropriate observation notes and data tables.
6. At your teacher's direction, collect your group's sample of swamp water. Prepare your sample for incubation, using the procedure detailed in Student Information 6.2 and Student Activity 6.3, Part D. Use the dilutions indicated by your teacher.

PART C. Counting the Coliform Colonies

7. The following day, as a group, follow the procedure detailed in Student Activity 6.3, Part E, for counting the coliform colonies.
8. Report the solution descriptions and results to your teacher and the class.
9. Collect and record each group's results in your data table.

Observations

DATA TABLE

Group	Solution Tested	Actual Fecal Coliform Count (colonies/100 mL)	Fecal Coliform Count Equivalent to Undiluted Sample (colonies/100 mL)

Analyses and Conclusions

1. Using the formula provided in Student Information 6.2, convert the results for each sample into the equivalent fecal coliform count for an undiluted sample.
2. Calculate and record the average colonies of fecal coliform per 100 mL of undiluted sample.
3. Write a description of the data, including comments on whether any or all are acceptable, and why. As appropriate, recalculate the average colonies, leaving out any data you argue are inaccurate.

Performance Criteria

- Work is done safely in the laboratory and good technique employed for preventing microbial contamination or infection.
- All members of the group follow the work plan.
- Group follows the procedures learned during the previous class sessions. Group performs the dilutions properly.
- Group uses a "control plate" as a sterility check for the equipment.
- Group collects and reports data in a proper data table.
- Group results compare favorably with those of others; if not, the group offers a rational explanation for any error or difference.
- Report is well-written, showing an understanding of the results provided by the other groups. It includes appropriate explanations and conclusions. It displays correct grammar and spelling and is written in paragraph form.

TEACHER NOTES

LESSON 7 Environmental Assessment

Focus

Students will learn about procedures required for obtaining a permit to conduct changes that affect any wetlands, rivers, or streams. They will use the results of surveys and tests performed during *Rivers Biology* to create an environmental assessment. They will practice community involvement through consideration of a specific proposed environmental change, and they will analyze the issues involved in such a process.

Learner Outcomes

Students will:

1. Learn about standards and regulations concerning water quality and the application of those rules under the direction of the Federal Clean Water Act, as administered by the U.S. Army Corps of Engineers, U.S. Environmental Protection Agency, and other state (or local) entities.
2. Use the results of their field-site study as part of the assessment of a proposed or potential environmental change that may affect that site or the local river or stream.
3. Learn how to use appropriate forms to conduct an environmental assessment.
4. Advertise and hold a public meeting concerning an environmental project.
5. Assess the impact on the local river or stream ecosystem of a real or simulated environmental change.

Time

Six or seven class periods of 40–50 minutes per period plus an optional field trip to a river or stream (one period may be an evening or weekend public meeting). Some work, such as research and meeting preparation, may be best performed as homework assignments. If time and resources allow, much of the activity may be done in class as group work, allowing extra time for research.

DAY 1: Student Information 7.1: Introduction to Environmental Assessment

FIELD TRIP: (OPTIONAL) Discuss proposal at site; if doing a new site, do all field-site tests and surveys.

DAY 2: Student Activity 7.2: Permit Request

DAY 3: Student Activity 7.3: Section 404 Permit Review

DAY 4: Student Activity 7.3: Section 404 Permit Review

DAY 5: Student Activity 7.4: Public Notice, Part A

DAY 6: Student Activity 7.4: Public Notice, Part B (includes public meeting) (may be evening or weekend meeting)

DAY 7: Student Assessment 7.5: Final Environmental Assessment and Statement of Findings

Safety and Waste Disposal

If students perform any laboratory or field work during this lesson, follow all safety procedures presented in Lesson 1 (and Lesson 6 if students test for fecal coliform).

Advance Preparation

Prepare to supply students with Student Information, Activity, and Assessment sheets 7.1 through 7.5. Gather all necessary equipment and materials. Arrange for a local expert to attend class and take part in the discussion of environmental assessment. Obtain current information from local water agencies on Clean Water Act regulations; they change frequently. Also inquire possible threatened or endangered species in the proposal area. If you are doing *Rivers Biology* in a country other than the United States, adapt this lesson to correspond with the regulatory processes pertinent to your area.

During an actual permit review, after public notice comes a 30-day comment period, announced in newspapers and on radio, during which anyone may submit facts or information about the project or voice arguments or objections. Decide whether you will assign "public comment," compiling this in an environmental assessment or environmental impact statement just as the Corps of Engineers does.

Decide what format the public meeting should use. You may allow presentations or moderate a debate. Debate works particularly well if you can incorporate help from a speech teacher. Decide whether students will give individual presentations from their own perspective or work together in groups, dividing different elements or points for presentation among themselves.

Decide what kind of audience should participate in the "public meeting." Choices include: other Rivers Project classes, other science classes, parents, environmental professionals, public officials, and the community. If the change is an actual proposal, students may find great enthusiasm and value in holding a real public meeting. Inviting students in classes such as Social Studies and English may add a different twist to the meeting. If you are assigning roles to audience members, make up role cards or name badges before the meeting. For more on role playing, see *Rivers Chemistry* and *Rivers Geography*.

As needed, make arrangements for auditorium space and for attendance by public officials and others. As appropriate, issue written or verbal invitations to specific individuals or groups you would like to have attend. (Alternately, you may plan to have students do this.)

If students will create publicity for the meeting, such as posters, press releases, or articles for the school paper, prepare materials and lesson plans for these activities. Allow sufficient days in the class schedule for the publicity to reach its intended audience before the meeting; alternatively, have students do Student Information 7.1 and prepare the publicity earlier in *Rivers Biology,* so the meeting can proceed directly from the class days spent on Lesson 7. For specific teaching activities on newspaper writing, see *Rivers Language Arts.*

This lesson is designed to focus on a real, potential, or simulated proposed change that may affect the local river or stream field-site the students have already visited and studied. Examples of such projects include: causeways, road fills, dams and dikes, property protection, reclamation devices such as rip rap, bike and walking paths, fills, beach protection, levees, sanitary landfills, and diversion of the water for agriculture or development. (Projects that involve wastewater discharge fall under a more elaborate permitting system than the Corps of Engineers 404 permit used in *Rivers Biology.* If you must use such a project as your proposed change, make sure students are aware that the actual permitting process is different.)

If you cannot determine any such change for your site, select another site for which such a change seems more likely or instructive for your class. If using a different site, students will need to repeat the earlier field-site surveys and tests. This will add significant class and field-trip time to the lesson.

Prepare to provide in the classroom or guide students in finding maps, newspaper articles, television programs, actual government reviews of the field site, interviews, and library or online research relevant to the field site and the proposed change.

Materials

Student Activity 7.2: Permit Request

Per student

journal

Student Information 7.1: Introduction to Environmental Assessment

Per group

topographic map of study site, 7.5- or 15-minute series

results from Student Activities 4.4, 4.5, 4.6, 5.4, 5.5, and 6.3

results from other Rivers Project classes, if available

public information about the proposed change

plant identification guide

animal identification guide

unlined paper

colored markers

Student Activity 7.3: Section 404 Permit Review

Per student

journal

Student Information 7.1: Introduction to Environmental Assessment

Per group

topographic map of study site, 7.5- or 15-minute series

results from Student Activities 4.4, 4.5, 4.6, 5.4, 5.5, 6.3, and 7.2

results from other Rivers Project classes, if available
public information about the site and the proposed change
plant identification guide
animal identification guide

Student Activity 7.4 Public Notice and Meeting

Per student
paper name badge
journal
3" × 5" cards or other note materials

Per group
black or blue marker
visual aids, such as flip chart or map

Optional
microphones and sound system
podium
video or slides of site under consideration
slide projector or videocassette player
projection surface
audiocassette recorder, connected to sound system
audiocassette tapes
video camera and tripod
videocassette
role-playing name cards for audience members

Student Assessment 7.5: Final Environmental Assessment and Statement of Findings

Per student
journal
Student Information 7.1: Introduction to Environmental Assessment
3" × 5" cards or other note materials prepared for public meeting
notes taken during public meeting

Per group
topographic map of study site, 7.5- or 15-minute series
results from Student Activities 4.4, 4.5, 4.6, 5.4, 5.5, 6.3, 7.2, 7.3, and 7.4
information from local water agencies on Clean Water Act regulations
results from other Rivers Project classes, if available
public information about the site and the proposed change
plant identification guide
animal identification guide

Vocabulary

Clean Water Act
cost-benefit analysis
dredged material
endangered
environmental assessment
fill material
mitigation
navigable capacity
navigation
National Pollution Discharge Elimination System (NPDES)
permit
public interest review
Section 404
threatened

Background for the Teacher

In the activities of Lesson 7, students conduct an environmental assessment of a real or simulated proposal for an activity that potentially impacts a wetlands site. The organization of the activities parallels the process undertaken by official governmental agencies. You can make this a simulated or realistic scenario, depending on your local circumstances and your classroom time and interests. If agencies in your area are actually considering a proposed change that may impact the environment of your local river, stream, or wetlands, you and your students can use this to inspire and guide your work in this concluding lesson in *Rivers Biology*.

Introducing the Lesson

1. Give students an overview of Lesson 7. Tell students about the chosen issue and the associated public meeting. If you prefer, have students propose and decide on a simulated or proposed change to evaluate.
   If you have not already done so on your own, decide with the class on the audience, date, time, and publicity.
2. Have the students read, discuss, and answer the questions for Student Information 7.1: Introduction to Environmental Assessment. (Answers for student sheets are in Appendix B.)
3. Have a guest speaker from one of the regulatory agencies discuss the permitting procedure and issues related to development that affects a wetlands or riverine site. Agencies involved in the 404 Permit process include not only the U.S. Army Corps of Engineers but the U.S. Fish and Wildlife Service, U.S. Environmental Protection Agency, state natural resource departments, soil and water conservation districts, and local agencies.
4. If students are assessing a proposed project at a site other than the one tested earlier, have the class perform the field-site tests and surveys from Lessons 4 through 6 on the new location. Allow for a lengthy field trip and additional lab work after returning from the field, some of which needs to be completed five days after the field trip. To reduce the field-trip length, divide the activities among small specialist groups, instead of having the entire class conduct each test or evaluation. Students will benefit from repeating the activities, and you may use these activities as an additional assessment tool.

5. Have students gather together all the results from their surveys and tests related to the local river or stream field site.

Developing the Lesson

1. Divide the class into two role-playing teams. One team is the Project Development Team; the other is the Reviewing Agency, such as the U.S. Army Corps of Engineers. You may divide the students into multiple smaller groups working independently as each type of team. For instance, you could assign two Project Development Teams, one acting environmentally responsibly, the other without environmental priorities.
2. Have students in a Project Development Team carry out Student Activity 7.2: Permit Request. Meanwhile, have the Reviewing Agency read that activity and work on publicity for the meeting. When the permit is completed, allow the Development Team to create publicity. You may limit student work on Student Activity 7.2 to one class period or allow extra time for online, library, or other research. Complete publicizing the meeting.
3. As homework, have students read Student Activity 7.3: Section 404 Permit Review.
4. Prepare photocopies of the Project Development Team's Permit Request. Distribute it to the Government Review Team (or give one to each member).
5. If students will go on a field trip during Lesson 7, have them meet as teams at the field site to discuss the proposed project.

6. Have students in the Government Review Team carry out Student Activity 7.3: Section 404 Permit Review, in which they determine whether or not the proposal complies with the Section 404 Guidelines.
7. Discuss the results of the permit request and review process. Encourage students to recognize that a process such as this, in which different groups have different interests, may reveal conflicting viewpoints. Have them consider differences in how various groups handled the process and its components. If the permit was not granted, or the proposed project found not to comply, what can each side do to facilitate compliance? Let them identify issues not resolvable through discussion or concession.
8. Have the students carry out Student Activity 7.4: Public Notice and Meeting, Part A, in which they prepare their presentations for the mock Public Interest Review to identify additional factors of aesthetics and general public interest. Have students do this preparation in class or as homework, as individuals or in groups. Emphasize the importance of active participation by each student, and specify how much time each speaker will have. Prepare the students for the experience of role playing. Encourage them to take a chance by stepping out from their individual beliefs and into some with which they may not agree.
9. Have the students carry out Student Activity 7.4: Public Notice and Meeting, Part B, Holding the Public Meeting. Make the meeting as professional as possible, with the audience for which you earlier arranged. Use either presentation or debate format. To make the meeting more realistic and ensure a variety of audience participation, give out role cards or name badges to audience members as they arrive, using such roles as citizens, representatives of environmental groups, landowners, elected officials, newspaper reporters, television reporters, and representatives of other interested groups. Remind students to stay in their roles for the duration of the meeting. Allow the audience time to ask questions and comment about the project.

Concluding the Lesson

1. Have all students, working in groups (or as individuals), play the role of Government Review Team to carry out Student Assessment 7.5: Final Environmental Assessment and Statement of Findings. Emphasize that students should use this assignment to demonstrate their understanding of stream assessment.
2. Direct students where to post their reports, such as in the classroom, on a school bulletin board, school newspaper, or even a more public location, depending on the circumstances.
3. If not already done, have students compare results of their tests and studies with their water-quality predictions they entered in their journals at the conclusion of Lesson 1.

Assessing the Lesson

1. As a class, discuss all the Final Environmental Assessment and Statement of Findings prepared by the class. Compare results, paying attention to similarities and discrepancies. Discuss the value of the environmental impact statement and the completion of this simulation. Here is a suggested scoring rubric, which you can use with the performance criteria included in the student assessment sheet.

Scoring Rubric

Score	Expectations
0	Group has written/done nothing.
1	The environmental assessment is minimal. Narrative does not cover all necessary topics. Information does not reflect the entire scope of available information. Findings are not based on actual assessment or inaccurately reflect data. Group has relied on others to provide information on conclusions. Group did not follow format requirements. Member contribution was very disproportionate.
2	The statement of findings is articulate, follows stated format, and draws reasonable conclusions. Omits some important information or consideration of relevant data. Shows an understanding of the basic concepts and factors in a Environmental Assessment and Statement of Findings. Does not sufficiently discuss alternatives. Member contribution was somewhat lopsided.
3	Member contribution was proportional, and members worked together responsibly. Findings are appropriate given the available data and draw reasonable conclusions. Thorough discussion of alternatives and their pros and cons. Well-articulated statement exhibits solid understanding of the elements of a Environmental Assessment and Statement of Findings.
4	Same as 3, but Summary of Findings is concise, highlighting major elements of the decision. Uses outside data sources and demonstrates high level of coordination and understanding. Proposal alternatives consider novel solutions to project problems. Demonstrates superior understanding of the process and has compelling conclusions.

2. If, in Lesson 7, students used a field site different from the one originally surveyed and tested for the earlier lessons in *Rivers Biology,* have them prepare a report comparing and contrasting the environmental quality of the two sites. Students must support their positions using the results of their observations and analyses.

3. Have students share their results with other Rivers Project classes, either in your school or via the Rivers Project telecommunications system. Decide whether students will use results of Student Assessment 7.5: Final Environmental Assessment and Statement of Findings, or of individual tests and surveys conducted at the field site during *Rivers Biology*. To share via telecommunications, have students do computer-oriented Challenge Project: Telecommunications (located in Appendix A). For more information on telecommunications, see page x in unit introduction and pages 210–213 in the Teacher Notes for the Challenge Project: Telecommunications.
4. If not already done, have students share their results with public officials. For specific teaching activities on political action writing, see *Rivers Language Arts*.
5. Have students do or complete their collage based on what they have learned during *Rivers Biology*, or specifically about their field site. (For Student Assessment: River of Your Dreams: A Collage and suggested scoring rubric, see Appendix A.)
6. Have students complete their portfolios or other unit-long assessment projects. (For Student Assessment: Portfolio and suggested scoring rubric, see Appendix A.)

Extending the Lesson

1. Have students make a collection of related newspaper and magazine articles related to environmental assessment. Post each to a classroom bulletin.
2. Visit meetings being held on local issues relating to development on a river-related site or other water-sensitive issue. To locate such meetings, contact the county clerk, city council, or state and federal environmental regulatory agencies, especially water-management authorities.
3. Have students repeat the necessary surveys, tests, and Lesson 7 activities required to conduct an environmental assessment of a different scenario. You may have students analyze: an alternative scenario for the same situation they have just assessed in this lesson; a proposal for change at a different site in the watershed of the local river or stream; a proposal for some change to the local watershed they believe is likely to be proposed in the future.
4. Have students prepare a report identifying, analyzing, and recommending some action that would benefit the local river or stream ecosystem. Indicate the extent to which students should base their work on the tests and surveys carried out as part of *Rivers Biology*.

Name

STUDENT INFORMATION 7.1

Introduction to Environmental Assessment

In the preceding lessons in *Rivers Biology,* you performed tests and surveys of your local river or stream and the surrounding environment. Your results have provided a more accurate understanding of the water and environmental quality of the river or stream and its surroundings. Scientists, researchers, and government agencies use such observations and analyses as part of the process of predicting the effects of potential changes to a watershed and its waters.

The Role of Environmental Assessment

Many companies, governmental entities, and individuals have reasons to fill, change, or drain wetlands. (Rivers, streams, and lakes are wetlands by scientific definition, but governmental regulations typically define them as surface waters.) In order to protect and preserve public waters and insure that the country's water resources are used in the best interest of the people, the U.S. Congress has passed laws requiring that any proposed alteration to a wetlands, river, or stream receive careful study by appropriate regulatory agencies in order to determine what negative effects it might have, before granting permits for such change. A **permit** is a document from a governing agency that grants permission to carry out some activity, in particular, an activity that may have ecological, social, or economic impacts.

Regulatory agencies charged with this permitting authority include the U.S. Army Corps of Engineers and the U.S. Environmental Protection Agency. If this study determines that the proposed action could have potentially negative effects, the agency conducts an **environmental assessment,** a procedure for determining the effects of a proposed activity on the environment (including the human environment). The agency ultimately uses the completed environmental assessment in order to determine whether or not to allow that specific activity.

DRIFTWOOD

In the 1990s, the number of environmental reports and publications informing the public increased tenfold.

Involvement of Federal and State Agencies

Water travel and shipping have been very important to national defense, so, since the 1890s, the U.S. Army Corps of Engineers has had the duty of keeping such activities functioning. The Corps take responsibility for protecting **navigation** (water transport by vessel) and the **navigable** (NAV ih guh bull) **capacity** of this nation's waters, meaning their accessibility for travel by ship or boat. This includes dredging channels and canals, building locks and dams, and regulating other activities that might threaten navigation, such as dam building, water diversion for agriculture and industry, and sand and gravel quarrying.

In 1968, Congress revised this mandate to require that review of permit applications involve the potential effects on fish and wildlife, conservation, pollution, aesthetics, ecology, and general public interest. This new type of review was identified as a **public interest review,** because it is supposed to weigh the comments and concerns of the public along with scientific studies and analyses. The review determines if the alteration is in the best interest of the people of the United States. Permits are granted or denied after considering all the effects of the proposed activity. The Corps denies a permit if the activity will threaten the health or lives of humans, or if it will destroy known populations of **endangered** (in immediate risk of extinction) or **threatened** species (declining in number throughout their range).

DRIFTWOOD

Effects of the management activities of the U.S. Army Corps of Engineers on the Missouri River: Channelization, dams, and dredging in the last 50 years have shortened the river by 130 miles; it is now one third its original width, 98 percent of the original islands and sandbars are gone, and one fifth of native species are endangered.

Source: American Rivers (www.amrivers.org/amrivers/)

Clean Water Act

In 1972 Congress enacted the Federal Water Pollution Control Act with the purpose of restoring and maintaining the chemical, physical, and biological integrity of this nation's waters. This act was amended in 1977 and renamed the **Clean Water Act.** Through the Clean Water Act, permits for activities potentially harmful to bodies of water became more numerous and strict. The Corps of Engineers permit program for **Section 404** of the Clean Water Act regulates physical changes to wetlands.

Section 402 of this act established a set of regulations called the **National Pollutant Discharge Elimination System (NPDES),** covering industrial and municipal source discharges of chemical, biological, and thermal pollutants into the nation's waters. The NPDES permit program is administered by a state agency or the U.S. Environmental Protection Agency (EPA). The NPDES permit requires that applicants meet the discharge limits and water-quality standards set by each state or by the EPA.

Amendments to this act authorized the Corps to regulate any discharge of materials into any waters of the United States. It requires permits for point-source discharge. A separate permit program regulates nonpoint source pollution, specifically pollutants in dredged or fill material. **Dredged material,** material removed from an underwater site (such as a river or bay bottom), can contain heavy metals or other contaminants. **Fill material,** material added to or removed from a land site, typically a construction site, can contain hydrocarbons or heavy metals such as mercury, lead, or cadmium.

The Environmental-Assessment Process

Different types of activities involve different permitting processes, but most proposals fall under the authority of the Corps of Engineers. The requesting group must submit a permit request that includes a description of what activities they plan to do, and when. When the District Office of the Corps receives the request, their staff studies the site, potentially affected surrounding areas, and the proposed change.

This process usually involves a general review of the proposal and field work. A Corps of Engineers representative, as well as personnel from other agencies (such as the U.S. Environmental Protection Agency, U.S. Fish and Wildlife Service, and the local soil and water conservation districts), talks with the permit applicant about the project and compliance with regulations.

If the project is particularly large or in an ecologically sensitive area, Corps staff do field work and data collection to gather more information. They survey the site for endangered and threatened species, critical breeding areas, and other phenomena the project might affect. The survey team may include experts in biology, earth science, archeology, geology, and chemistry. The team may even perform a **cost-benefit analysis** of the project in the local economy, comparing the economic gains and losses of this project with those of other alternatives. They look for any information that might enable them to make the best decision.

The reviewers establish a 30-day comment period, during which the public can declare their concerns or support for the project. The permitting agency may hold a public meeting to determine public reaction and impact.

Using the gathered data, the team evaluates the environmental impacts of the proposed change. Based on these results, they make a determination of requirements. They may decide to permit the proposal instantly, if the proposed project has no possible negative effect or is a common, popularly accepted practice and does not violate any laws. They may find the proposal not in compliance for any of the reasons listed in Table 7-1.

TABLE 7-1

Section 404 Criteria for Denying Permit Request
1. A less damaging practicable alternative can be found;
2. The action will result in significant degradation of the aquatic ecosystem;
3. The action does not include all practicable and appropriate measures to minimize potential harm to the aquatic ecosystem;
4. The action violates state water-quality standards;
5. The action jeopardizes existence of threatened or endangered species; or
6. The action violates a federally designated water sanctuary.

Although the Corps denies some requests, they grant a permit if a project will have minimal negative effects or positive impacts outweigh negative. The Corps may notify the requesting party of actions required for compliance

DRIFTWOOD

Dams on the Columbia and Snake Rivers are responsible for the elimination of more than 200 salmon runs and place 76 additional runs in jeopardy of extinction. The operation of these dams also eliminates the spring freshet, which flush young salmon to the sea, greatly increasing the length of time it takes juvenile salmon to migrate to the sea. Though installed fish ladders accommodate the passage of adults upstream, no changes have been made to meet the needs of smolts traveling downstream.

with the Clean Water Act. The requesting group may need to change the proposed activity in order to reduce its environmental, social, or economic impact. If the project will fill, drain, or permanently destroy a wetlands, the developer must provide **mitigation,** some compensatory action, often in another location, to offset those negative effects. The mitigation may require restoring another wetlands area, creating a protected, undeveloped wetlands 33 percent larger than what will be destroyed. Presently, most permits are approved.

Questions

1. Explain why federal regulations require an entity that plans a development involving aquatic ecosystems to obtain a Section 404 permit.
2. What other factors besides navigation does the U.S. Army Corps of Engineers review during Section 404 permitting?
3. What is the purpose of the Federal Water Pollution Control Act?
4. In your own words, list the steps in an environmental assessment.

Name

STUDENT
ACTIVITY 7.2

Permit Request

Purpose

To understand some elements of making decisions about how to best use our nation's waters, and to use knowledge gained in past lessons to help determine the potential effects of human development on your local river or stream.

Background

Your group will act as a Project Development Team, representing a developer who plans to carry out the change is that the focus of your study for your local river or stream. You will research the project, its site, and its projected effect on the ecosystem. Based on these results, you will prepare a permit request, then submit it to the Government Review Team. Because your team wants the Review Team to permit this change, you must determine the likely effects of your planned activity in order to develop the most acceptable proposal.

The standards by which the Government Review team completes its evaluation are the Section 404 guidelines; the reasons for not granting a permit are listed in Table 7-1 in Student Information 7.1. The Review Team may grant the permit, grant it on the condition that you alter your plans, or deny it because the alteration is too environmentally detrimental. Provide information to support the permit and assure its acceptance.

Materials

Per student

- journal
- Student Information 7.1: Introduction to Environmental Assessment

Per group

- topographic map of study site, 7.5- or 15-minute series
- results from Student Activities 4.4, 4.5, 4.6, 5.4, 5.5, and 6.3
- results from other Rivers Project classes, if available
- public information about the proposed change
- plant identification guide
- animal identification guide
- unlined paper
- colored markers

Procedure

1. If your teacher has not done so, select one group member as project manager. This person is in charge, keeps the group on task, clarifies questions with the teacher, and makes sure the group and its members perform responsibilities for the permit request.
2. As a team, provide the applicant information listed in the Observations section.
3. As a team, determine and describe the location and scope of the proposed project, including its purpose and need, as listed in the Observations section. (Use the location identification technique your teacher indicates.)
4. Determine the environmental impact of the proposed project. This includes effects on navigation, ecology, and water resources (both public water supply and water conservation). Use test and survey results to identify the current conditions at your site. Use such data, plus information from outside

sources, to hypothesize how the proposed change will affect those variables. If your teacher so instructs, form into pairs to handle specific sources such as maps, newspaper articles, television programs, actual government reviews of this site, interviews, and library or online research. Focus on the project and its impact on the aquatic ecosystem. Identify and list all possible effects of the proposed change. Include any negative effects on such factors as water quality, water availability, habitats, and wildlife. Consider not only end results but also effects of any related construction. Record in your Observations section how and why you think the proposed change may or may not affect these factors.

5. Using the same kinds of resources as for Step 4, determine the cultural effects of the proposed project. Analyze whether it would destroy or degrade an archeological or cultural site. Evaluate the extent to which it is consistent with traditional uses of the land. Record in your Observations section.
6. Using the same kinds of resources as for Step 4, determine how the proposed project would affect such social concerns as recreation, employment, the local economy, public safety, and overall quality of life. Be sure to consider whether the project will protect people or jeopardize their safety. Record in your Observations section.
7. Using the same kinds of resources as for Step 4, determine and record any other concerns raised by this proposed project. Record in your Observations section.
8. Make drawings or diagrams that help clarify the project and its potential impacts.
9. If your group thinks that the Government Review team may find that your proposed project does not comply with the Section 404 guidelines listed in Table 7-1 of Student Information 7.1, work as a group to consider mitigations or alterations to the proposed project you could offer in order to increase the probability of obtaining a permit.

Observations

PERMIT REQUEST

Applicant: Name ____________________ Date ____________________

Address ____________________ City, State, Zip ____________

I. Description and Scope of the Proposed Project

A. Project Location ____

River or Stream ____

Site Address ____________________ Nearest Town ____________

B. Project Description

C. Purpose and Need of the Proposed Project

II. Environmental Impact of the Proposed Action and Any Alternatives

A. Navigation

B. Ecological Concerns

- C. Water Resources
  1. Public water supply
  2. Water conservation

III. Cultural Value
- A. Historical and Archaeological
- B. Other

IV. Social Concerns
- A. Recreational
- B. Public Safety
- C. Economy and Employment

V. Other Concerns

Analyses and Conclusions

1. Use your completed observations to create a permit request that follows the headings in the Observations section. Your group must determine what is important to include. Distribute a copy of the completed request form to each member of the review team.

Critical Thinking Questions

1. What portion of the permit application was the most time-consuming? Why?
2. If you were the permitting authority, what decision would you make about this permit request? Why?
3. Comment on the extent to which you think your point of view as a project developer influenced what you wrote on the permit application.

Keeping Your Journal

1. Speaking as the project developer, why should this project receive a permit?
2. Without regard to your assigned role in this activity, what is your personal opinion about this proposed project?

Name

STUDENT ACTIVITY 7.3

Section 404 Permit Review

Purpose

To understand some elements of making decisions about how to best use water resources, and to use knowledge gained in past lessons to help determine the potential effects of human development on a selected site.

Background

Your group will act as a Government Review Team, representing a public agency, such as the U.S. Army Corps of Engineers, charged with reviewing applications for a permit to engage in an activity that may change the environment or navigability of your local river or stream. The developer will give you a permit application that provides information about the proposed project, its site, and its projected effects on the environment and community.

In a real review, applicants meet on site to discuss regulation compliance with a Corps representative and with staff from other agencies, such as U.S. Fish and Wildlife Service, U.S. Environmental Protection Agency, state or local soil and water conservation district, and the state department of natural resources. Whether or not your teacher has your team revisit the site with the project development team, you will investigate the site and the project. Your team must evaluate this information to determine the likely effects of the project and whether it complies with Section 404 guidelines.

Based on those determinations, your team will: issue a permit; make conditions for compliance, such as mitigation or modifications; or deny the permit for one of the reasons in Table 7-1 in Student Information 7.1, such as a less harmful alternative exists or degradation of the aquatic ecosystem will be significant.

Materials

Per student

- journal
- Student Information 7.1: Introduction to Environmental Assessment

Per group

- topographic map of study site, 7.5- or 15-minute series
- results from Student Activities 4.4, 4.5, 4.6, 5.4, 5.5, 6.3, and 7.2
- results from other Rivers Project classes, if available
- public information about the site and the proposed change
- plant identification guide
- animal identification guide

Procedure

1. If your teacher has not done so, select one group member as project manager. This person is in charge, keeps the group on task, clarifies questions with the teacher, and makes sure the group and its members perform responsibilities for the permit request.
2. If your teacher so instructs, form into pairs to handle specific resources such as maps, newspaper articles, television programs, actual government reviews of this site, interviews, and library

or online research to review the permit request form. Focus on the project and its impact on the aquatic ecosystem.

3. Note the background information about the site, such as location and permit applicant, as noted in the Evaluation Sheet in the Observations section.
4. Review the permit request carefully. Based on the results of your surveys and tests of the local river or stream, and on any other data available to your group, review the extent to which the proposed activity complies with Section 404. As a team, consider whether any alternatives to the proposal would be less detrimental. Consider whether it violates state water-quality standards, might jeopardize threatened or endangered species, or violates any federal designated water sanctuary (meaning an area set aside for the use of wildlife, such as a state or federal park, wildlife refuge, or game preserve). Evaluate whether the proposed project takes appropriate steps to minimize environmental impacts on the area. Mark Section I of your Evaluation Sheet accordingly.
5. Based on what you know about the area's current characteristics, and using the same kinds of resources as for Steps 2 and 4, evaluate whether the proposed activity will significantly affect the physical and chemical characteristics, biological characteristics, special aquatic sites, or human-use characteristics of the river or stream. Consider how the construction process will affect the site. Mark Section II of your Evaluation Sheet accordingly.
6. Consider the physical characteristics of the site that affect its suitability for the project and its susceptibility to negative effects. These factors include: depth of water; degree of turbulence; water velocity, direction, and variability; and soil characteristics. As a team, determine which factors should be considered and which are not relevant. You may obtain much of this information from your results of Student Activity 4.5: Completing a Riverine Habitat Survey. On Section III of your Evaluation Sheet, indicate which factors you considered in your evaluation, and the results of that consideration. Indicate whether the physical site is suitable for the proposed project.
7. Based on your observations and analysis so far, make factual determinations of the potential for short- or long-term adverse effects from this project. Mark Section IV of your Evaluation Sheet accordingly.
8. In Section V, note those responsible for preparing the evaluation.
9. Based on your evaluation, make your findings about whether the proposed site complies with the Section 404 guidelines. Enter these in Section VI. As appropriate, attach a statement that includes conditions for compliance or provides specific reasons why the proposed site does not comply with the guidelines.
10. When the evaluation is complete, have the project manager and all members of the Project Development Team sign Section VII.
11. Distribute copies of the completed review to the project development team, and discuss with them any recommended project revisions or mitigation.

Observations

Applicant: Name ______________________ Date ______________________

Address ____________________ City, State, Zip _____________

Description and Scope of the Proposed Project

Project Location ____

River or Stream ____

Site Address ______________________ Nearest Town ___________

I. REVIEW OF COMPLIANCE

Review of the permit application indicates that:

1. The activity represents the least environmentally damaging practicable alternative. yes __ no __

2. The activity does not appear to (1) violate applicable state water-quality standards. (2) jeopardize the existence of listed endangered or threatened species, (3) violate any federal designated water sanctuary. yes __ no __

3. The activity will not cause or contribute to degradation of U.S. waters, including adverse effects on human health, life stages of organisms dependent on aquatic ecosystems, diversity, productivity, and stability, or recreational, aesthetic, and economic values. yes __ no __

4. Appropriate and practicable steps have been taken to minimize potential adverse impacts on the aquatic ecosystems. yes __ no __

II. TECHNICAL EVALUATION FACTORS OF THE AQUATIC ECOSYSTEM

Characteristics	**N/A**	**Not Significant**	**Significant**
A. Physical and Chemical Characteristics	_______	_______	_______
1. Substrate impacts	_______	_______	_______
2. Suspended particulates	_______	_______	_______
3. Alteration of current patterns	_______	_______	_______
4. Alteration of water circulation	_______	_______	_______

Characteristics	N/A	Not Significant	Significant
B. Biological Characteristics			
1. Effect on endangered species and habitat	______	______	______
2. Effect on aquatic food webs	______	______	______
3. Effect on other wildlife (animals, plants)	______	______	______
C. Special Aquatic Sites	______	______	______
1. Sanctuaries and refuges	______	______	______
2. Wetlands	______	______	______
3. Mud flats	______	______	______
D. Human Use Characteristics	______	______	______
1. Effect on municipal and private supplies	______	______	______
2. Effect on recreational and commercial fishing	______	______	______
3. Aesthetic impacts	______	______	______

REMARKS: When a check is placed under the SIGNIFICANT category, add explanations on an attached sheet.

III. SITE DELINEATION

A. The following factors, as appropriate, have been considered in this evaluation:

1. Depth of water at site yes __ no __
2. Degrees of turbulence yes __ no __
3. Water velocity, direction, and variability yes __ no __
4. Soil characteristics yes __ no __
5. Other factors affecting the site yes __ no __

B. An evaluation of the preceding appropriate factors indicates that the site is acceptable for the action listed. yes __ no __

IV. FACTUAL DETERMINATION

Review of the appropriate information as identified in items above indicates that there is minimal potential for short- or long-term adverse effects as related to:

1. Physical substrate yes __ no __
2. Water circulation yes __ no __
3. Aquatic ecosystem structure and function yes __ no __
4. Cumulative impacts on the aquatic ecosystem yes __ no __
5. Secondary impacts on the aquatic ecosystem yes __ no __

V. EVALUATION RESPONSIBILITY

A. This evaluation was prepared by:

NAME	POSITION	DATE
____________	____________	____________
____________	____________	____________
____________	____________	____________
____________	____________	____________

VI. FINDINGS

A. The proposed site complies with the Section 404 guidelines. yes __ no __

B. The proposed site complies with the Section 404 guidelines with the inclusion of the following conditions. (See attached.) yes __ no __

C. The proposed site does NOT comply with the Section 404 guidelines for the following reasons: yes __ no __

1. A less-damaging practicable alternative can be found. yes __ no __
2. The proposed action will result in significant degradation of the aquatic ecosystem. yes __ no __
3. The proposed action does not include all practicable and appropriate measures to minimize potential harm to the aquatic ecosystem. yes __ no __

VII. SIGNATURES OF APPLICANT

1. ______________________________ Date __________
2. ______________________________ Date __________
3. ______________________________ Date __________
4. ______________________________ Date __________
5. ______________________________ Date __________
6. ______________________________ Date __________

Analyses and Conclusions

1. Based on your evaluation, does this proposal comply with Section 404 guidelines? Why or why not?
2. If the permit does not comply with guidelines, what suggestions would you make to the project development team to assist them in getting a permit?

Critical Thinking Questions

1. If you were making a real-life decision about this proposed action, what additional information would you want in order to make a final decision?
2. Does compliance always mean a proposal is harmless? Describe two situations in which a proposal may comply with guidelines but be environmentally detrimental.
3. Under what conditions might a proposal that does not meet all guidelines be acceptable?
4. What kinds of individuals and organizations are most likely to support this proposal? What would their main arguments be?
5. What kinds of individuals and organizations are most likely to oppose this proposal? What would their main arguments be?

Keeping Your Journal

1. What is your personal opinion about this proposed project?
2. Less than 40 years ago, these procedures and regulations did not exist. You may live in a house or attend school in a building that would now need a 404 permit in order to be built. Do you think these regulations have improved society, or have they made things too complicated? Support your answer.

Name

STUDENT ACTIVITY 7.4

Public Notice and Meeting

Purpose

To participate in a meeting similar to those that take place in actual permit review situations, articulating what you know about a particular river or stream environment in a practical context.

Background

After a permit proposal has been submitted, the review agency establishes a written comment period (usually 30 days) during which anyone can write to the agency with facts, arguments, or objections. The agency informs the public of the proposed activity through mailings, newspaper notices, and fliers in public facilities. The agency may hold one or more public meetings, usually only if someone specifically requests a meeting and identifies a significant issue that would warrant additional public review. The audience at a public meeting may include local businesspeople, environmentalists, concerned citizens, government agency staff, and students.

The meeting may begin with a presentation from the Corps of Engineers about the project and its suitability to the site. Then the development team presents their proposal. Then the public may address the review team with information or concerns, and ask the development and review teams clarifying questions.

The government review team uses any new information gathered at the meeting to make a final decision, weighing also the public's general attitude or feelings about the proposal. Toward such ends, you will hold a public meeting about the proposed project affecting your river or stream.

Materials

Per student

- paper name badge
- journal
- 3" × 5" cards or other note materials

Per group

- black or blue marker
- visual aids, such as flip chart or map

Optional

- microphones and sound system
- podium
- video or slides of site under consideration
- slide projector or videocassette player
- projection surface
- audiocassette recorder, connected to sound system
- audiocassette tapes
- video camera and tripod
- videocassette
- role-playing name cards for audience members

Procedure

PART A. Preparing for the Public Meeting

1. Determine the points you and your group would like to make regarding this proposal, keeping in mind available presentation time. Focus on your strongest points. Organize your material. As instructed by your teacher, prepare note cards or a complete statement to aid your presentation.
2. Create visual aids such as maps, charts, or slides that will help your presentation. Your teacher may provide additional guidance about formats.

3. Practice your presentation to make sure you are comfortable with the material and can complete your presentation in the allotted time.

PART B. Holding the Public Meeting

4. Prepare the meeting space. Use a sound system so everyone can be heard. If possible, arrange to audiotape or videotape the proceeding for future reference.
5. Bring your Student Activity 7.4 and journal to the meeting.
6. On your name badge, use the marker to write in large letters your name and the name of the group you will represent during the public meeting.
7. When the moderator begins the meeting, be sure everyone knows that this public interest meeting is not an official meeting by a review agency. Clarify whether the proposed project is an actual proposal or a simulation.
8. In your journal, using the format in the Observations section of this activity, make notes of each presenters main points. In particular, write down remarks that add information to, or refute, your argument. Be sure to record public comments and questions.
9. When it is your turn, speak from the perspective of your group (such as developer, landowner, or government reviewer). Watch your time, so the moderator does not have to cut you off before you have finished.

Observations

Date ______________________ Project Location ____________________

Brief Project Description ______________________________________

Key Points of Presentations

Pro: Con:

Public Interest Review

Public Concerns and Questions

Public Comments

In Favor: Opposing:

Analyses and Conclusions

1. Write a paragraph or two for each group, summarizing the main arguments they presented during the meeting.
2. Which presentation did you find most persuasive? Why?
3. Which presentation did you find least persuasive? Why?
4. Did any audience members bring up issues you had previously not considered? Identify and discuss them.
5. Based on your observations of the public's comments and behavior at the meeting, offer your opinion of the predominant viewpoint of the public toward the proposed project.

Critical Thinking Questions

1. In what ways are your current opinions about the proposed project different from those you held before the meeting?
2. To what extent were you able to consider information valid regardless of how it was presented? What information did you find difficult to accept as valid, and why?
3. If you were on a development team: Based on the information presented in the public meeting, how complete was the written permit request submitted by your team (as part of Student Activity 7.2)? What additional information should your group have included?
4. If you were on a government review team: Based on the information presented in the public meeting, how complete was the Section 404 Evaluation prepared by your team (as part of Student Activity 7.3)? What additional information should your group have included?

Keeping Your Journal

1. What did you like and dislike about public speaking?
2. If you were going do your presentation again for another audience, what, if anything, would you do differently?
3. Imagine you are working for the local newspaper. Write a newspaper story on the meeting.

Name

STUDENT ASSESSMENT 7.5

Final Environmental Assessment and Statement of Findings

Introduction

Using the accumulated data from preceding activities in Lesson 7, each group (or individual, as directly by your teacher) will now act as the Government Review Team to prepare its own final report that includes the issuance or the nonissuance of the Project Permit. Remember, all decisions and requirements must comply with Section 404 of the Clean Water Act.

In addition to ecological and human safeguards considered in doing the Section 404 Evaluation (in Student Activity 7.3), this final statement of findings carefully weighs the public reaction from the public meeting and from any written comments. Such input may mean the difference between granting or denying a permit.

This final determination also takes into account the requirements of regulations that protect historical and cultural sites from degradation or destruction. In a real review, the state historical society would provide a review of such issues.

Most of the critical issues in your decision-making process will derive from your knowledge of the river or stream and the potential effects of the proposed project. Reviewing data you have collected may provide sufficient information to allow you to make an accurate determination. In some cases, however, you may not have or be able to gather certain data, given your time and resource limits. Real reviewers face somewhat parallel limits, because they do not have the resources to consider every factor fully before making a decision.

Materials

Per student

- journal
- Student Information 7.1: Introduction to Environmental Assessment
- 3" × 5" cards or other note materials prepared for public meeting
- notes taken during public meeting

Per group

- topographic map of study site, 7.5- or 15-minute series
- results from Student Activities 4.4, 4.5, 4.6, 5.4, 5.5, 6.3, 7.2, 7.3, and 7.4
- information from local water agencies on Clean Water Act regulations
- results from other Rivers Projects classes, if available
- public information about the site and the proposed change
- plant identification guide
- animal identification guide

Procedure

1. List the description and scope of the proposed project, as indicated the Observations section of your assessment.
2. The Public Interest Review, Section II, is a broad summary of general public reaction to the proposal. Sometimes government agencies perform an in-depth study to analyze public interest. In most cases, however, the Public interest Review is a summation of public comments, reactions, and opinions. Use information gained from the public meeting to develop this section.
3. Using available resources, investigate and consider

whether the proposed project would harm any historical landmarks or other protected cultural sites. If time allows, contact your local or state historical society. List your conclusions as if you were the State Historical Review.

4. Section IV is usually a presentation and analysis of comments submitted to the reviewing agency by interested individuals and organizations. If your process does not include such input, skip this section.
5. From the surveys, drawings, and tests performed at the site, and from any other available data, determine all ecological, physical, land-use, and aesthetic concerns associated with the proposed project. To determine whether any species at the site are endangered or threatened, consult plant and animal identification guides. Based on data taken from the site, plus any other relevant information available, also determine the potential effect on water resources. You will have to determine the effects on navigation, cultural values, social concerns, and the economy relatively subjectively, such as through group consensus.
6. Make sure your assessment addresses any relevant information or comments from the public meeting.
7. When you, or your group if working together, have gathered all the analyses and made your decisions, prepare an environmental assessment and statement of findings, following the model presented in Observations.
8. Determine whether a permit is to be issued or not, with a notice of any actions developers must take in order to maintain the minimum adverse impact on the site area.
9. Develop a statement supporting your decision.
10. When the findings have been completed, make sure that both the project manager and your teacher sign it.
11. Reviewing agencies must post each report on permit approval or disapproval properly (usually in the local post office). Post your report as directed by your teacher.

Observations

ENVIRONMENTAL ASSESSMENT AND STATEMENT OF FINDINGS

I. DESCRIPTION AND SCOPE OF PROPOSED PROJECT

A. Applicant: Name ____________ Date ____________

Address ____________ City, State, Zip ____________

B. Description and Scope of the Proposed Project

Project Location ____________

River or Stream ____________

Site Address ____________ Nearest Town ____________

C. Project Description ____________

D. Purpose and Need of the Proposed Project ____________

II. PUBLIC INTEREST REVIEW

III. STATE HISTORICAL SOCIETY REVIEW

IV. COMMENTS BY INDIVIDUALS OR ORGANIZED GROUPS

V. ENVIRONMENTAL IMPACT OF PROPOSED ACTION AND ANY ALTERNATIVES

A. Navigation

1. Commercial navigation
2. Recreational navigation

B. Ecological Concerns

1. Fish and wildlife
2. Shellfish
3. Benthic macroinvertebrates
4. Endangered or threatened species
5. Wetlands

C. Water Resources

1. Public water supply
2. Water conservation
3. Water quality
4. Pathogenic organisms

D. Aesthetics

E. Cultural Value

1. Historical and archaeological
2. Other

F. Land Use

1. Conservation
2. Prime and unique farmlands
3. Food production

G. Economic Concerns

1. Energy needs
2. Employment
3. Growth

H. Social Concerns

1. Recreation
2. Safety

I. Physical Concerns

1. Erosion

2. Flood hazards

J. Alternatives Considered

K. Summary of Findings

Date ______________________ Preparer ______________________

Date ______________________ Reviewer ______________________

Date ______________________ Approving Official ______________

Performance Criteria

The successful environmental assessment and statement of findings will:

- Incorporate data on the local river or stream collected during *Rivers Biology* and from the Public Interest Review.
- Contain complete entries relevant to the subject of each portion of the report.
- Be written in clear, well-written statements.
- Include substantial relevant data and information gathered from appropriate outside sources.
- Make a reasonable finding, based on the content and result of the environ mental assessment.
- Discuss the pros, cons, and practicality of project alternatives.
- Follow format and content requirements provided by the teacher.
- Reflect active participation and valuable contribution of all members of the group, if assigned as a group project.

Appendix A

Assessments

This appendix contains the unit-long and unit-end assessments for *Rivers Biology*. (Assessment activities for specific lessons are contained within those lessons.) From the assessment tools presented in *Rivers Biology,* utilize what suits your class, curriculum, and preferences. Feel free to use your own assessment tools, performance criteria, and scoring rubric.

Teacher Notes for River of Your Dreams: A Collage (pages 214–215)

Focus

Each student constructs a collage that represents his or her feelings, gives a message, or shows concern for a river or stream. The collage is developed throughout the project and should include material that represents each of the lessons covered.

You may display the collages during the project or at an open house for parents and the community. The partially finished posters may be displayed during the unit so students may share ideas on making collages. You may also have students present their posters to the class using short (five-minute) presentations.

Materials

Per student
poster board
large envelope for holding collected materials
scissors
markers
glue stick
Per class, optional, will vary
magazines from which pictures and text can be cut
colored paper
computer
Internet access
color printer

Advance Preparation

Prepare to supply students with Student Assessment: River of Your Dreams: A Collage (pages 214–215). Decide whether students will have class time for this assignment or should do it at home. Collect all equipment and materials necessary for the work students will do in class. If possible, arrange for students to have access to the Internet as they search for pictures to use. Many good river pictures are available on the World Wide Web that students can print with a color printer.

Make arrangements to display the completed posters, such as in the classroom, in hallways, in school display cases, or at special events. Decide whether to have students deliver oral presentations in conjunction with their completed posters.

You may wish to invite your school art teacher or local artists to talk to your students about designing a collage. Ask for guidelines they would use to evaluate artistic merit. Invite them to view the finished projects.

Developing the Lesson

1. Have students read Student Assessment: River of Your Dreams: A Collage for ideas and suggestions about making a collage. Explain that they will work on their collages during each lesson segment of the *Rivers Biology* unit. Suggest that, in the early lessons, students make sketches of their collage, starting to glue materials onto their posters only after covering a number of lessons and making at least one trip to the local river or stream. Students may want to wait until a few weeks before the assignment is due to begin the final assembly. Make sure they understand that they will have to add ideas for each lesson in the unit. Indicate what materials are available for their collages. Give minimum and maximum sizes for posters.
2. Making a collage is an unusual type of assessment for most science classes. Emphasize to your students that you will evaluate the collages like all assignments, so they require careful planning and execution.
3. Inform students whether the posters will be displayed throughout the project and where the completed posters will be displayed.
4. If students will do oral presentations, provide other appropriate guidance on the presentations, including length.
5. As the completion date approaches, arrange for publicity for the posters and presentations as appropriate, such as contacting school and local newspapers, sending notes home to parents, and so forth. You may have students participate in these activities.
6. You may want to collect a copy of the abstracts to compile a book that visitors and parents can use to interpret the displays. Having the students organize the display like an art show makes the effort seem more realistic and should produce better results.
7. Performance criteria are included in the student handout. Here is a suggested scoring rubric.

Scoring Rubric

Score	Expectations
0	No collage is completed.
1	A collage was started but is incomplete or shows no theme or message.

2	A collage is completed and a theme or message used. Only a few of the lessons are represented by the collage. Questions are completed.
3	A collage is completed and a theme or message evident. Materials from most lessons were included and mostly supported the theme. The summary abstract was completed and some understanding of the themes and ideas presented in the collage was shown.
4	A collage is completed that conveys a theme or message using ideas from all the lessons. The display is neatly constructed and shows good use of space. Responses written in the summary or abstract show an understanding of the use of collages to convey a message.

Teacher Notes for Portfolio (pages 216–217)

Focus

Each student selects and arranges multiple items that demonstrate the students learning of river-related scientific content, attitude, and skills.

Materials

Per student
manila folder
plastic report cover or three-ring notebook

Advance Preparation

Look through the *Rivers Biology* lessons. The Teacher Notes sections give suggestions on using the Student Activity sheets, Student Assessment sheets, and additional projects for assessing and extending the lesson. Select the activities you will assign, making sure students will have a variety of opportunities to express their abilities and interests during this river study.

Based on your assignment plans, decide what kinds of items students may (or must) include in their portfolios, and modify Student Assessment: Portfolio (pages 216–217). If you have used portfolio assessments, select good examples to share with your students. If not, select some ideas from different lessons to make a sample portfolio.

As the *Rivers Biology* unit is undertaken, students should save what they plan to include in the portfolios. Ideally, they should keep all selections in manila folders in the classroom.

Developing the Lesson

1. Have students read the handout, Student Assessment: Portfolio. Discuss the variety of work students may select for their portfolio, and show a sample portfolio that includes a variety of items such as special reports, laboratory work, writings, and group reports. You may want to supply a list of required items and list of suggested items.

2. Give each student a manila folder or envelope and show where in the classroom the folders will be kept.
3. Show how students should prepare the portfolio for final submission, such as in a bound notebook, folder, or a three-ring notebook.
4. Give target dates for assembling the various portfolio items and for final submission.
5. Performance criteria are supplied in the student handout, Student Assessment: Portfolio. Here is a suggested scoring rubric.

Scoring Rubric

Score	Expectations
0	The portfolio has not been collected and documented. Only random papers are included, and no summary statement is written.
1	A portfolio has been collected, but work has not been selected to reflect student performance; work is incomplete; portfolio does not have introduction, afterword, or other reflective analysis.
2	The notebook is organized and mostly complete. A limited understanding of biology concepts is shown. Writings are logical but often incomplete.
3	The notebook is organized and complete. A satisfactory understanding of concepts and applications is shown. Writings are logical, often including additional observations and readings.
4	The notebook is organized and complete. Biology concepts are understood and readily applied to analyze data and draw conclusions. Additional observations, readings, and resources are frequently cited.

Teacher Notes for Biological Field-Study Proficiency Checklist (pages 218–219)

Focus

As students do the activities in *Rivers Biology,* you may want to track and evaluate their proficiency on specific field-work and lab skills. During *Rivers Biology,* students will use many of the skills listed in the table on page 209.

Skill proficiency is a basic goal of many *Rivers Biology* activities, so you may want to use checks to build student proficiency at a particular task before evaluating it. On a check card, you may indicate evaluation of the skill without grading proficiency. Students who come to your class with good skills in microscope use will demonstrate proficiency immediately. Others may have to be taught how to work with the microscope and use it several times, with feedback, before being ready for a graded assessment proficiency.

Materials

Per student

Biological Field-Study Proficiency Checklist

Per class

materials and equipment used during laboratory and field activities

For the teacher

laminated copy of checklist

clipboard

marking pencil to write on laminated checklist

permanent checklist in computer or grade book to transfer items evaluated

Advance Preparation

Select the *Rivers Biology* activities you will assign. Adapt Student Assessment: Biological Field-Study Proficiency Checklist (pages 218–219) to reflect the proficiencies you will evaluate. If students will also do activities from other Rivers Curriculum Guides, such as pH or turbidity tests from *Rivers Chemistry,* add them to the checklist (page 209). Prepare to distribute that checklist and the student assessment handout to each student. Laminate a copy of the checklist for your own use. You may also want to prepare proficiency cards that indicate when you have made an evaluation.

Developing the Lesson

1. Have students read the handout, Student Assessment: Biological Field-Study Proficiency Checklist and the checklist. Discuss the work and field laboratory proficiencies that they will be taught in contact and expected to master.
2. Students can demonstrate that they know how to use the materials, identify the organisms, or conduct the test by arranging for an observation, by the teacher's casual observation in the field or classroom, or during a formal laboratory test. Explain how you will do evaluations.
3. If desired, give target dates for each evaluation and for checklist completion.
4. During the course of *Rivers Biology,* complete the checklist for each student. Use a laminated copy for working in the field, though you may transfer notations to a permanent copy.
5. Performance criteria are supplied in Student Assessment: Biological Field-Study Proficiency Checklist. Here is a suggested scoring rubric.

Scoring Rubric

Score	**Expectations**
0	Demonstrated only a few proficiencies, or many were inadequately completed.
1	Attempted each proficiency item, but most work reflects poor performance or inadequate attention to the activity and its completion.
2	Achieved proficiency on each proficiency item, and work reflects average student performance. Work demonstrates a

limited understanding of laboratory procedures and concepts. Written analyses are logical but skimpy or incomplete.

3	Achieved proficiency on each proficiency item, and work reflects excellent student performance in all areas evaluated. Student has followed laboratory procedures and attempted to analyze data and draw conclusions. Observations and notations are logical, often including additional observations and readings. Student helps others meet proficiency.
4	Achieved proficiency on each proficiency item, and work reflects superior student performance in all areas evaluated. Student understood laboratory procedures and readily applied to analyze data and draw conclusions. Written work frequently includes additional observations, readings, and resources. Student helps others meet proficiency.

Biological Field-Study Proficiency Checklist

Mark the number that best describes the proficiency.

Poor = 1, Average = 2, Good = 3, Excellent = 4, Superior = 5

Student Name

Skill Type	**Lesson**	**Proficiency**										
Equipment Use	1, 4	Compass										
	2, 4	Microscope										
	4	Dip net										
	4	Kick net										
	4	Insect net										
Water Tests	4	Stream velocity										
	4	Temperature										
	6	Fecal coliform										
Chemical Tests	5	Dissolved oxygen										
	5	Biochemical oxygen demand (BOD)										
Specimen Identification	3, 4	Benthic macroinvertebrates										
Other	4, 5, 6	Data sheets										
	4	Habitat assessment										

Teacher Notes for Challenge Project: Telecommunications (pages 220–222)

Focus

In this culminating activity, students use data from other river or stream sites to expand their knowledge of water quality in rivers and streams. The activity emphasizes the importance of analyzing collected data, looking at results over time and distance, and making comparisons. Once students have retrieved river data and information electronically, they will interpret and report those data and share the results with a web site location and fellow students.

Time

Three or four class periods of 40–50 minutes, with additional homework time for preparing presentation and making telecommunications contact.

Materials

Per group
computer
telecommunications system
list of river-related Web sites
map of North America or the world, with rivers
maps of rivers, streams, and watersheds being studied
computer disks
Optional
spreadsheet program for sorting and managing data
graphics software
graphing software

Advance Preparation

Gather all necessary equipment and materials. Create a list of river-related Web sites. A partial list is included in Resources, Appendix C, though more sites are coming online daily. (Information on accessing the Rivers Project via e-mail and the World Wide Web is included on page x. *Southern Illinois University at Edwardsville,* the *Office of Science and Mathematics Education,* and *Rivers Project* are key words to finding the Rivers Project home page for information on using the database, for asking questions about the project, and for connecting with other schools testing their local river or stream.) Make these connections yourself, to make sure you provide clear, accurate, and current addresses and instructions. As necessary, make a handout with details on how to use the school's telecommunications system. Prepare to supply students with a telecommunications handout, a list of Web sites, and Student Assessment Challenge Project: Telecommunications (pages 220–222).

If your class shares computers with others in the school, make arrangements for your class to have appropriate access. Arrange for an e-mail address (and perhaps Web site) students can use for communicating with the web sites where they are making river comparisons. Work with your school's computer resource person to give students proper preparation and guidance for the telecommunications portion of this activity.

If students will make group presentations, decide how the group and the individuals will be evaluated. Decide how students will present their investigations to you and the class, such as written reports, oral presentations, Web pages, or poster presentations. Prepare a handout of presentation guidelines if different from what is included on the student sheet.

Introducing the Lesson

1. Have students read and discuss Student Assessment Challenge Project: Telecommunications. Distribute the list of Web sites.
2. Explain that the students will organize the data collected during the past lessons in *Rivers Biology* and compare the river they tested with some other river. The group must select a river and find data on the Web. Using data and information gathered there, they will develop an organized report comparing the two sites, make contact with a person who will receive this comparison, and finally telecommunicate the package for possible publication at the Web site.
3. Explain school or classroom computer capabilities and class access to such facilities. Distribute and clarify any telecommunications handout. Ascertain which students have experience communicating via e-mail and the World Wide Web and have access at home. Take the students to visit the computer room when possible.
4. Describe the form in which students will present the results of their research. Distribute handouts of presentation guidelines as appropriate, with deadlines.

Developing the Lesson

5. Divide the class into groups of 4–6 to work on this project, trying to distribute students with computer networking skills evenly among the groups. Each group should choose a specific river, watershed, or geographic area to investigate. As a group, have them explore the Web to see what rivers have data and information available. If information is hard to find, have the group select a different river.
6. Working within the group, have students divide the individual research that fits within the group's general area of interest.
7. Have the groups retrieve from the Internet the observations and data required for each student's individual research endeavor. Peer coaching in using the system should help students with little background in the technology.
8. While some students work with the computers, have others investigate their test sites using maps, geography books, and other library resources.
9. Encourage students to graph their data. For specific teaching activities on graphing, see *Rivers Mathematics*.

Concluding the Lesson

10. Have students prepare a printed (or electronic) copy of the transmittable report for your evaluation. Have them include a copy of communication made with the affiliated Web site.
11. Have the students send the report to the appropriate Web site.
12. If desired, have groups prepare and deliver oral presentations to the class on the transmitted work.

Assessing the Lesson

The challenge project can take many forms, so assessment criteria should fit the specific tasks involved. In most cases, the assessment criteria can be the same you would generally use for written analysis and for group participation. Criteria for the telecommunications portion of the assessment will depend on student access to and experience with telecommunications systems. For certain activities, you may want specific scoring rubrics. Here are some possibilities:

Scoring Rubric for Telecommunications Presentation

Score	Expectations
0	Nothing is produced.
1	Narrative is minimal, has little relevance to biological study and water-quality tests or standards, or does not describe the work correctly. The group relied on others outside their group to retrieve information from Web sites. Collected and tabulated little local or Web site information.
2	Members adequately organize the data and information from the local field site and make some connections between the test results and site conditions. At least one member manipulated the telecommunications system. Collected and tabulated information comparing the local river site with the site for which information was downloaded. Includes some charts and graphs. With some structural or grammatical revision, the final report is ready for transmission.
3	Students meaningfully organize the data and information from the local field site and make some connections between the test results and site conditions. Most members manipulated the telecommunications system. Collected and tabulated some information comparing the local river with the site for which information was downloaded. Charts and graphs make information easier to understand. Made comparisons in a logical sequence and information satisfactorily summarized. Cited appropriate references. Final report is ready for transmission.
4	Fulfills the requirements for 3 and has reasonable explanations or correlation among test data. They have researched water

testing beyond requirements for the biological theme, or otherwise show interest and scholarship in environmental concerns. All members accessed the telecommunications system with minimal help.

Scoring Rubric for Group Presentation

Score	Expectations
0	No poster/presentation is made.
1	The poster/presentation shows little preparation or has incorrect or meaningless information.
2	Students followed the criteria for preparing the poster and have some correct and meaningful information. The presentation is the result of efforts of most, if not all, of the group. Presentation reflects the areas investigated by most group members. One element (either poster or presentation) may be good, but the other is poor.
3	Both the poster and presentation meet most of the criteria. All group members participate. Presentation clearly reflects the areas investigated by each group member.
4	Work meets all the criteria. Shows exceptional effort by all members for both poster and presentation.

River of Your Dreams: A Collage

Introduction

When you try to explain a situation or feeling to others, often a picture works better than words. A collection of pictures may be even more effective. A collage is a collection of pictures and other materials, selected and arranged to present a message to the viewer. Over the course of the *Rivers Biology* unit, you will construct a collage whose message relates to a river's water quality, the organisms living in or along its banks, and the uses of rivers and streams by humans.

In the early stages of your collage, you will need to decide what message you want to convey. Your message should involve each of the lessons you will study in the *Rivers Biology* unit, but it should not just be a unit summary or jumble of water and river pictures. In considering what message you want to communicate, you may want to consider the quality of the river site you are depicting. What living things can be found along or in that river? How are humans affecting the site either positively or negatively? What things are taking place at the site? What parameters affect the site the most? Do you hike or picnic along its shore? Is it used for your drinking water? Does treated sewage water end up there? What industries rely on the river?

Decide what you want others to be more aware of. You may want to consider the beauty of the river or stream; its value as habitat; its value to humans as a water supply, a source of transportation, or a recreational resource; its power and destructiveness in a flood; or the effects of such human actions as removal of vegetation from stream banks, construction of dams, drainage of wetlands, dredging, pollution, or overuse by industries or municipalities. Will it be a good river, a bad river, a river of the future, or a river of the past?

Materials

Per student

- poster board
- large envelope for holding collected materials
- scissors
- markers
- glue stick

Per class, optional, will vary

- magazines from which pictures and text can be cut
- colored paper
- computer
- Internet access
- color printer

Procedure

1. Decide on a general topic or theme. Write out the message you want to convey. List kinds of pictures you could use to convey that message. Decide if you will need to include other materials or verbal messages. Make sketches of possible arrangements.
2. After completing each lesson, add to your collage. Determine a message from that lesson that adds to the central theme. Your message should weave through the different lessons. As you select and gather materials, you may find pictures in magazines or at various water-related Internet sites. Drawings, even cartoons, might help. Each time you consider materials for your collage, ask yourself, How will this express my message? If you have no answer, look for different materials instead.

3. Collect your collage materials in a large envelope labeled with your name and class. Keep the envelope in the classroom (as directed by your teacher) until you attach them to your poster. Do not attach materials to your board until you have covered several lessons. Especially, wait until you have made a trip or two to your local river or stream.
4. Experiment with the sizes, shapes, and colors you have collected. Anticipate what you might want to add after upcoming lessons. The arrangement should be visually pleasing, with thought given to color, form, shape, and line. Balancing pictures, print, and white space contributes to the effectiveness of a collage.
5. When you are satisfied with your arrangements, attach the materials to the poster board. Include a title in the collage that represents your theme. The title can be included in the collage or added to the display as one would in an art gallery.
6. When your collage is completed, tape a sheet of paper on the back with your name, class, and date. Also write a paragraph that abstracts or summarizes your collage. Include responses to the instructions that follow.
   - Provide your name, and your theme or topic.
   - Describe the message you wished to convey.
   - Describe how you have used pictures to convey that message.
   - Add any other information that will help the reader make more sense of the collage and better understand its message.

Performance Criteria

You have artistic freedom in your use of pictures, printed materials, and other items for your collage. Your grade will be based on whether you have followed the procedures described in this sheet. The successful collage will:

- Feature a single message or theme throughout the collage. The message is evident through pictures and organization and may also be stated in print.
- Contain materials representing each *Rivers Biology* lesson studied.
- Clearly demonstrate organization and planning. The collage is neatly constructed.
- Reflect artistic considerations, such as use of color, space, balance, and line.
- Have an accompanying sheet containing complete responses to the instructions listed in Step 6 of the student handout. Responses show insights into the use of a collage to present a message.
- Be complemented by an oral presentation (if assigned) that emphasizes the message and the relationship of the different parts of the collage to the message.

Portfolio

Introduction

A portfolio is a collection of the work you deem the best and most representative. In preparing your portfolio, include materials you would be proud to show others and you would keep. The purpose of a portfolio is to observe your own progress and to share what you have learned with others. A well-thought-out portfolio will allow you, classmates, and teachers to assess what you have learned and done.

Procedure

1. During this study of *Rivers Biology,* you will keep your best papers or products in a manila folder in the classroom. Update the contents of this folder regularly, so that it represents your best work. The items may include a compilation of entries from your journal, laboratory or class reports, homework, drawings, poems or stories, artwork, photographs, articles written for a newspaper or published journal, or any other evidence of the effectiveness or benefits of this project.
2. You may also include articles you have found in newspapers, books, or magazines regarding environmental issues (such as water quality and industrial accidents), national or international water issues, or comments and happenings related to the local river or stream. Include your own observations or comments about the significance of the event or any impact on the community for each article included.
3. Organize the portfolio into sections that represent the lessons or areas being presented. Mount the materials or punch as necessary before placing them in the sections. Supply a table of contents.
4. Complete the portfolio by writing an overview of your involvement in *Rivers Biology.* The overview should include a brief introduction, a reflection on what you have learned and done, and a concluding statement. Address the writing to an adult who has no knowledge of this project, but whom you want to impress positively.

   In *Introduction,* include information about the project and state the source or assignments for the works included in the portfolio.

   In *Reflections,* answer the following questions:
   - What did you learn or discover during this project? What skills did you gain?
   - What part of the project was most important? What was most fun? What part was most useful?
   - In what ways have you become more aware of your environment during this project? Have your attitudes changed?

   In *Conclusion,* reflect on where you have been and where you might go because of your interest in the environment in general, in rivers and streams specifically.
5. Complete your portfolio by assembling all materials into a binder, such as a plastic report holder or a three-ring notebook. Place your materials in the following order:
   - Title page, including your name, class, and date
   - Table of contents
   - Project overview
   - Your collection of works organized in some way

Materials

Per student

- manila folder
- plastic report cover or three-ring notebook

Performance Criteria

Your portfolio will be evaluated at the end of the *Rivers Biology* unit. Your grade will be based on whether you have followed the procedures described in this sheet. The material in your portfolio should reflect the following:

- Acquisition of new knowledge and skills.
- Ability to express ideas clearly and to think critically.
- Ability to analyze data and draw conclusions.
- Ability to revise and expand your understanding.
- Descriptions of the problem-solving process as well as solutions to problems.
- Connection of information from several disciplines.
- Self-reflection and evaluation.

Biological Field-Study Proficiency Checklist

Introduction

As you do the activities in *Rivers Biology,* you will gain proficiency at many specific skills related to laboratory and field work. Your teacher will provide you with a list of such skills. You will be asked to display proficiency in each of the skills. Your teacher will explain how and when that proficiency will be measured.

Procedure

1. Place a copy of the Biological Field-Study Proficiency Checklist in your journal or notebook for ongoing reference. You can also copy the proficiencies in your journal and any expected dates for completion.
2. After completing each lesson, determine whether you are proficient at using the materials and equipment to complete the included procedures. If not, or if you are not sure, arrange to work with the materials until you achieve proficiency.
3. Some proficiency evaluation can occur only in the field; some takes place in the laboratory. You will be expected to work with fellow group members to gain each proficiency. Your teacher will be ready to provide assistance.
4. Most of the listed skills are hands-on. Missing either field or laboratory sessions may have a negative impact on your ability to develop these skills.
5. On your list, track which skills you have been evaluated on and which you have mastered. If the evaluation is not final, then continue to work toward proficiency.

Materials

Per student
- Biological Field-Study Proficiency Checklist

Per class
- materials and equipment used during laboratory and field activities

For the teacher
- laminated copy of checklist
- clipboard
- marking pencil to write on laminated checklist
- permanent checklist in computer or grade book to transfer items evaluated

Performance Criteria

Your teacher will evaluate your performance with each of the listed tasks or proficiencies when doing field or laboratory work. Your grade will be based on how well you perform each task. Each task will be evaluated independently. Important points to consider are:

- Have you completed work with the listed skill and demonstrated proficiency?
- Can you relate the test or material to the *Rivers Biology* lesson studied?
- Can you clearly demonstrate skills at using the instrument or material in field or laboratory settings?
- Can you explain the uses and procedures for the specific test or material?
- Have you worked with other members in your group to make sure that all are proficient at the task?

Telecommunications

Introduction

In this project, you will use the Internet to find sites from which to gather information about rivers and with which to share the data you have collected about your own river or stream site. You will probably have significant independence in selecting what types of data to gather, what types of comparisons to make, and how to display your results and analyses. This is your opportunity to figure out how you want to apply the results of your *Rivers Biology* tests and surveys to river understanding and stewardship.

You will work in groups or as individuals as directed by your teacher. You will use the data gathered during *Rivers Biology* to develop a profile of your river or stream to share with others. You will prepare tables and graphs of the data, and you may collect maps or additional information from the library or the Internet. After you interpret and organize this information, you will write (and perhaps upload to the Internet) a report comparing your river or stream site with another river or site. Finally, you may work in groups to develop a poster presentation to share with your class the results of your telecommunications-based comparison.

Materials

Per group

- computer
- telecommunications system
- list of river-related Web sites
- map of North America or the world, with rivers
- maps of rivers, streams, and watersheds being studied
- computer disks

Optional

- spreadsheet program for sorting and managing data
- graphics software
- graphing software

Procedure

PART A: Using Information from the World Wide Web

1. Look at a map of North America or the world. As a group, identify a river you can compare with your local river or stream. You may want to compare two rivers you think are similar, such as the Columbia and the Nile. You may want to contrast two rivers you think are quite different, such as the Snake and the Illinois, or the Hudson and the American. You may want to use two rivers that drain into the same larger body of water.
2. On the World Wide Web, locate organizations or Web sites that contain information about that river, particularly data about its water quality. If you cannot find useful data about that river, select a relevant river for which you can find information.
3. Collect data from that Web site about the selected river. Look for data on topics for which you have collected data about your own river or stream. Search for other sites to collect data and assemble additional information.

PART B: Providing Information for Sharing with the Computer Network

4. On the Web, locate organizations or Web sites that contain information about your local

river or stream or about other rivers in that region. Contact the site directly, such as by sending an e-mail message from the Web site (or contacting the webmaster). Inquire whether they would be interested in receiving the data you have collected, or a report that compares data from your river with that from another river. Save copies of all e-mail correspondence.

5. If you cannot locate an organization or site with information about your local river or stream, you may ask the river-related site you identified in Part A whether they would like data comparing their river or stream with yours.
6. Review the data about your local river or stream and site that your class has collected during *Rivers Biology*. Decide what data would be useful to share with other organizations or schools.
7. Review the data available from the Web on the river chosen for comparison. Determine what information you can compare directly. If information is missing, determine whether you can find it from another source. For instance, if someone at another site has flood data, but your class has not collected such information, you may be able to locate flood data for your local river or stream on a state Web site and integrate it into your own report.
8. Each member of your group should take responsibility for some specific part of the report. Develop an outline for your report. Be sure to consider not only the data you are most interested in comparing but basic information that affects the significance of those comparisons, such as season of the year, weather conditions, and temperature.
9. Gather precise location information, such as coordinates, for your report. To do this, locate your local field site on maps, and look at surveys or other available resources.
10. Develop a scheme for presenting test data and comparing test results. Consider which types of graph or charts would best represent your data.
11. Analyze your results, and prepare a presentation of your findings. Include comparison of all relevant data available. Make sure the report and each element it contains have clear titles. Include a contact person (and e-mail address) for comments.
12. Prepare your material for transmission. If you create individual documents on separate computers, work as a group to combine them. Input your group report onto the school computer that has telecommunications capability, and save the file or files for transmission.
13. Print a copy of your report, and submit it to your teacher. Indicate the Web site to which you plan to submit your report, and include copies of all e-mail communication about this submission.
14. When your teacher has approved your report for transmission, follow your teacher's instructions for uploading the data to the approved Web site.
15. If directed by your teacher, work as a group to prepare an oral presentation that shares with your class the results of your report.

Performance Criteria

A written report should:

- Demonstrate that you can access data electronically.
- Show that you understand the biological and water-quality parameters studied.
- Show that you have researched and performed critical thinking in evaluating the relationship between biological criteria and water quality of two locations.
- Include suggestions for improving or maintaining the water quality as it relates to organisms living in or on a river.
- Reference resources used in preparing the report.
- Compare and contrast your site with the one you have studied for the report.
- When combined with all the group's components into the final file or files, demonstrate the participation of everyone in the group.

A group oral presentation should:

- Represent the efforts of everyone in the group.
- Have a graphic display such as data, maps, charts, or graphs with accompanying explanations, sufficiently large to be clearly visible to the audience.
- Include an oral presentation in which everyone in the group participates.
- Be well-organized and easy to follow.
- Include an opportunity for the audience to ask questions; all members of the group should respond to questions.

APPENDIX B

Answers and Solutions for Student Information and Student Activity Sheets

Lesson 1: Answers to Questions

Student Information 1.1
Questions

1–3. Answers will vary.

Student Information 1.2
Questions

1. Answers will vary. Examples: "Ol' Man River" (*Showboat,* a musical), "Shenandoah River," "Across the Wide Missouri," "Red River Valley," "The River" (Garth Brooks), "River in the Rain" (*Big River,* a play), "Just Around the Riverbend" (*Pocahontas*), "River of Dreams" (Billy Joel).
2. Answers will vary. Examples: *Old Man and the Sea* (Ernest Hemingway), *Life on the Mississippi* (Mark Twain), *Showboat* (Edna Ferber), *Big Muddy* (B.C. Hall and C. T. Wood).
3. Answers will vary. People record important events in their lives, interesting ideas they have had, and things they have learned.
4. Answers will vary. People's diaries have been used by historians, anthropologists, sociologists, and even natural scientists to uncover information from the past. They are often the only records of local events.

Student Information 1.3
Questions

1. She performed an experiment without permission; she was not wearing safety gloves; she improperly disposed of the acid; she took off her goggles before leaving the lab; she opened the Petri dish with her hands; she did not wash her hands when finished; she did not clean her work area.
2. The experiment could have had a very dangerous result; chemicals could get on her hands, causing burns; the acid could corrode the plumbing or contribute to pollution; the acid could have splashed or glass could shatter in her eyes, causing injury or blindness; she could contract a coliform infection, destroy the validity of the coliform study, or both; if the acid were on her hands, she could subject others to acid burns; if she had coliform contamination, she could subject others to bacterial infection.
3. Jack did not stay with the group; he did not stay in the designated area; Jack climbed on a log, which could have been slippery.
4. Jack could have gotten lost or injured and no one would have known; he could have entered a dangerous area such as a poison ivy patch; logs are often very slippery and easy to break; by climbing out over the water while he was alone, he put himself in a very dangerous situation.

Student Information 1.4
Questions

1. Only groundwater and fresh surface water, less than 1% of the Earth's water.
2. Surface water is primarily replaced by precipitation and groundwater discharge.
3. Answers will vary. Humans must realize that water resources are not infinite. Only a limited amount of water is available and must, therefore, be carefully monitored.
4. a. condensation; b. evaporation; c. ocean; d. transpiration; e. precipitation; f. surface water; g. runoff; h. groundwater; i. infiltration.
5. Answers should place pollution in precipitation, runoff, infiltration, lakes, and rivers; may also include evaporation by air pollution.

Student Activity 1.5
Observations

1–7. Answer varies by state.

8. Answer varies by region.
9. Answer varies by state but usually includes bacterial contamination, agricultural runoff and fertilizer, and sedimentation.

10. Answer varies by state. Streams around major populations are usually less clean.
11. Answer varies by area. Private wells, city well, treated water, rivers, lakes, reservoirs, bottled water.
12. Answer varies by area. Some may go to a septic system; some is treated by a municipal plant; some may go directly into a body of water.

Analysis and Conclusions

1–2. Answer varies by state and region. Rivers in mountains and on steep gradients have straight flow, while those through flatlands meander.

Critical Thinking Questions

1. Answers will vary.
2. Answer varies by area but may include flooding, increased topsoil erosion. Low areas are more affected by increases in rain.
3. Answer varies by area. Some streams maintained primarily by groundwater discharge flow even through regional droughts, while others flow only from runoff.

Lesson 2: Answers to Questions

Student Information 2.1
Questions

1. Food chains consist of linear series of interactions between organisms. Each organism is a link in the chain, eating one thing and being eaten by another. A food web, on the other hand, is made of many chains, forming complex interactions among many organisms.
2. Most energy ultimately comes from the sun. This energy is converted to usable energy by plants through the process of photosynthesis. Whenever an organism is eaten, the energy is passed on to the consumer.
3. Primary consumers, sometimes called herbivores, are organisms that consume or get energy from producers, the organisms, like plants, that make solar energy usable. Secondary consumers, carnivores, eat other consumers.
4. Structures A and B would collapse. C would not.
5. C would endure. Because the web is not linear, the columns would be weakened but not to the point at which it would break, unless more pressure were placed on the web from the outside.

Student Activity 2.2
Analysis and Conclusions

1. The plants should be the most numerous. They provide the basis of the food web and provide needed oxygen for the other organisms.
2. The oxygen would become depleted, the water would be fouled up, and food would be consumed faster than it could be replenished.
3. The very small organisms are the smallest food sources, and consume and recycle waste material, capture energy, and release nutrients.
4. Answers will vary. If absent, then there are too many fish, the invertebrates are unable to reproduce, or the water was sterilized somehow. If present, then they are able to freely reproduce and are not under population stress by their predators.
5. Answers will vary.

Critical Thinking Questions

1. Answers should reflect the aquarium's problems.
2. Answers may include: The diversity is much lower; there are fewer niches; it is shallower; it receives light from all sides.

Student Activity 2.3
Analysis and Conclusions

1. 18–20 chains can be found.
2. Algae has the most connections, because it is the base of many food chains. Eel grass is close, as it is eaten by many primary consumers.
3. Some organisms are in more chains than others, but none can be in all. (Even the eagle is unlikely to eat a raccoon or a heron.) Some are in only a few. If a species disappeared, all the links it formed would disappear, breaking those chains. The absence of one species might not destroy the web, but it will significantly weaken it.

Critical Thinking Questions

1. The one with the most links.
2. Answers will vary, but students could insert a fungus, an earthworm, bacteria, or similar organism.

Student Activity 2.4
Analyses and Conclusions

1. Answers will vary. All the winners are dependent on luck. If one appeared more often, it would be random chance.
2. Answers will vary, but practice makes memory, and students should have learned the chains in this game.

Critical Thinking Questions

1. Answers will vary.
2. The higher up the food chain, the less impact the organism has. If the consumer is a factor that controls another organism, then its loss allows the other to reproduce more rapidly. Usually, organisms at the top of the chain have a variety of food sources and are a part of many chains.
3. Producers are algae and eel grass. Primary consumers are daphnia, snail, beetle larva, and slough darter. Secondary consumers are fathead minnow, white crappie, Johnny darter, small-mouth bass, red-eared sunfish, large-mouth bass, and rock bass. Tertiary consumers are great blue heron, raccoon, eagle, and large-mouth bass.
4. Answers will vary. For example, a brassy minnow could eat a small-mouth bass. The supporting argument might be that the bass was very young and small.

Lesson 3: Answers to Questions

Student Information 3.1
Questions

1. Answers will vary but may include the amount of pollution or pollution sources, the amount and quality of habitat, and the health of the organisms living there.
2. Answers will vary. Researchers try to be neutral by objectively selecting factors to consider. A stream assessment, for example, may rate a stream on qualities important to humans, not necessarily the inhabitants. Some factors deemed bad by humans in an index may turn out to be neutral or not as bad as they were rated.
3. Answers will vary. For instance, sun index, heat index, and wind chill were designed to meet the needs of weather news consumers.
4. Answers will vary.

Student Information 3.2
Questions

1. An index is designed to evaluate a situation using a variety of factors. These factors are combined in a meaningful way to give the scientist, researcher, or consumer a single number for making comparisons or for analysis.
2. Low diversity suggests an absence of more specialized organisms. Low diversity may mean that many specialized animals cannot live there because of their need for specific conditions. A healthy stream has a variety of conditions, including variations in temperature, velocity, surfaces, and other species that create many habitats that can, in turn, support different kinds of organisms. When species diversity is low, the food web is much less complicated, making it even more susceptible to change. Loss of diversity may indicate pollution, which could have affected those habitats.
3. A diversity index describes the species composition of a river. The pollution tolerance index, on the other hand, assesses the proportion of pollution-intolerant invertebrates present to those tolerant of pollution.
4. The number of individuals present in a given area, or density, gives an idea of how much life is present in an area and, thus, how much food and habitat are available. The number of species, or richness, gives an idea of how diverse the sample is. The proportion of each taxon, or evenness, gives the researcher an idea of how important each group is to the entire community.

Student Information 3.3
Questions

1. A clear, clean, oxygen-rich stream, with virtually no pollution will most likely hold Group I macroinvertebrates.
2. Nutrient enrichment, often regarded as pollution, increases the populations of many species that feed on algae and decaying material. (Some species, like the left-hand snail, are adapted to low-oxygen environments.) The lack of diversity, and competition, in a polluted environment favors those species that can exist, leaving them free to eat and multiply.

3. The riffles of streams contain the most macroinvertebrates.
4. Answers will vary depending on local river or stream.
5. Group I: Pollution intolerant; very little evidence of pollution; no artificial nutrient enrichment. Group II: Moderately pollution intolerant; natural forces such as stagnation of water and increase in temperature may occur seasonally; some nutrient enrichment is probable. Group III: Fairly pollution tolerant; low oxygen and fairly high nutrient enrichment. Group IV: Pollution tolerant; high nutrient enrichment; may also have high sedimentation and turbidity.

Student Activity 3.4
Analyses and Conclusions

1. The WQI is rated good with a value of 79.
2. The Q-value is a way of turning raw data into data in the same format. You cannot add values that have two different units of measurements. So, water scientists created a way to convert many types of water-quality data into the same type of value, a Q-value.
3. The WQI is found by multiplying a Q-value by its weighting factor. Not only does the WQI involve converting test results to a common type of value, it also weights results from some types of tests more heavily than others to come up with a total value for water quality.

Critical Thinking Questions

1. Fecal coliform, because it may be a direct threat to human health (though students may not know this, and answer dissolved oxygen, because it is weighted more heavily; if so, explain that the main issue in the WQI is not human beings but the overall aquatic habitat.)
2. Students may select dissolved oxygen because it is most heavily weighted; aquatic animals must have certain levels of dissolved oxygen in order to respire.
3. Answers will vary. Chemical tests support the biological tests in most situations. Low species diversity may occur for unknown chemical reasons. A recent introduction of a chemical or a periodic introduction of a chemical may kill off a portion of a community.

Lesson 4: Answers to Questions

Student Information 4.1
Questions

1. A diverse community tends to be a healthy community.
2. Answers will vary but may include: dams for electricity or water supply, levees for flood control, rip-rap for stabilization, channelization for transportation, and diversion for irrigation.
3. Answers will vary. Physical changes may include temperature, sedimentation, bank erosion, and trash. Chemical changes may include nutrification from agricultural runoff, decrease in dissolved oxygen, increase or decrease in pH, and introduction of toxic substances.
4. Answers will vary. Familiarity with the site will enhance planning for the field work and increase the ease of data collection. Knowledge of the habitats at the site helps determines sampling techniques and equipment. Familiarity and planning also make the experience safer and more enjoyable.

Student Information 4.2
Questions

1. Answers will vary according to site. For a rocky-bottomed riffle, sample a 1-m^2 area with a kick net and scrape all large rocks for invertebrates. For a muddy-bottom stream, sample vegetation, debris, and bottom with hands and dip-nets.
2. Collect fish first. Return soon after sampling. Sample from downstream to upstream. Sample debris, roots, snags, and bank first, then submerged vegetation and finally the substrate. Replace all materials removed from the habitat.
3. To minimize the disturbance of the habitat to be sampled.
4. Drawings should include dip-net, kick net, field guides, and collection trays.

Student Information 4.3
Questions

1. Answers will vary.
2. The presence of submerged plants indicates low turbidity and low stream velocity. Excessive algal growth suggests an overabundance of nutrients in the water.

3–4. Answers will vary depending on site.
5. Answers will vary. Pollution evidence must be from a general (nonpoint) source, like runoff.

Student Activity 4.4
Analyses and Conclusions

1–2. Answers dependent on data.
3. Diversity typically is greatest in riffles, but answer depends on local situation.
4. Answers will vary.

Critical Thinking Questions

1–3. Answers dependent on data.
4. The stonefly, alderfly, dobsonfly, and snipefly taxa would increase the index level.
5. Answers will vary.

Student Activity 4.5
Analyses and Conclusions

1–6. Answers will vary.

Critical Thinking Questions

1–3. Answers will vary.

Student Activity 4.6
Analyses and Conclusions

1. Answer depends on results.
2. Answer depends on results. The more taxa present in the sample, the higher the diversity will be.
3. Answer depends on results. Some taxa are more tolerant of the environment than others.
4. Answer depends on results. The density of each taxon in relation to the entire sample is called evenness.

Critical Thinking Questions

1. They should be consistent. The index values measure the same organisms; therefore, they should register ups and downs in the same direction.
2. Answers will vary.

Lesson 5: Answers to Questions

Student Information 5.1
Questions

1. Dissolved oxygen is the amount of oxygen present in a given amount of water. Dissolved oxygen is important because it is required by animals for respiration.
2. Warmer water holds less gas per unit volume.
3. Answer will vary according to area. Trout areas, for instance, are typified by cool, highly oxygenated streams. Catfish and carp thrive in water that is much warmer and much lower in oxygen.
4. Answers will vary. If there is a long-term change in temperature, nutrient load, or percent saturation, other fish species may become more common.
5. The oxygen in the pond may have been depleted by bacteria, algae, or other organisms, or the temperature may have risen to the point at which the water could no longer hold sufficient oxygen for the fish. Dissolved oxygen tests should show low saturation for the both scenarios.
6. The deeper water, because it is lower in temperature, holds more oxygen and is, therefore, more preferable for fish. This could be tested by testing the dissolved oxygen at both depths.

Student Information 5.2
Questions

1. Dissolved oxygen is the amount of oxygen in a water sample under natural conditions. BOD is the amount of oxygen consumed (demanded) by microorganisms during a certain amount of time. BOD indicates how much DO is actually available for aquatic organisms.
2. If BOD is high, less oxygen is available for macroorganisms, so organisms more tolerant of low oxygen migrate or reproduce in this environment, while those that need oxygen migrate away or die.
3. Answers will vary. Composting is the best alternative for yard waste. Clippings should not be dumped into the pond, because they reduce oxygen.
4. Heavy rains wash nutrients and debris into waterways, increasing the BOD.
5. BOD cannot be determined without first testing the DO. Knowing DO means little if the amount used in decomposition is not known. If the water quality was high, one would expect high DO values and low BOD.

Student Activity 5.3
Analyses and Conclusions

1. Answers will vary, but the coolest samples that have been agitated the most have the largest amounts of oxygen.

2. Answers will vary. Some samples may have more oxygen but be less saturated because of their temperature.
3. The oxygen is removed in the distillation process.
4. The amount of oxygen would decrease if the temperature increased, microorganisms were present, or both. Oxygen can exchange between the air and the water, moving in both directions.
5. The aeration process would dissolve more oxygen into the water. Colder water holds more dissolved oxygen.
6. Dissolved oxygen can vary even at one site. Multiple samples help researchers prevent sampling errors. If a large discrepancy occurs, the water can be retested.
7. Answers will vary. Differences may be caused by temperature, sampling error, and aeration technique.

Critical Thinking Questions

1. Answers will vary. Shaking is equivalent to rapids or falls, swirling is equivalent to riffles or wave action, mixing with a utensil is like water moving past a downed tree, and no aeration is like a stagnant pool.
2. Wave action supplies oxygen to lake environments. In small lakes or ponds, adding a mechanical aerator can improve water quality.
3. In environments with a lot of fine sediment or low flow velocities. Ponds with an abundance of algae and other plants often have high oxygen levels. When the plants die, their decomposition requires oxygen.
4. Answers will vary. High turbulence, rapids: darters, stoneflies, and mayflies. Medium turbulence, riffles: trout, darters, algae, and riffle beetles. Low turbulence, runs: small-mouth bass, craneflies, and eel grass. No turbulence: bluegill, catfish, and algae.

Student Activity 5.4
Analyses and Conclusions

1. Answers will vary. Low dissolved oxygen or low percent saturation leads to low Q-values and suggests poor water quality.
2. Answers will vary. Variations may be caused by sampling error, collection technique, or temperature difference. If one sample was taken deeper than another, it may have a lower DO due to a lack of photosynthesis, higher decomposition activity, less agitation and mixing, lower temperature, or a combination of such factors.
3. Answers will vary.
4. Answers will vary. Debris, turbidity, low stream velocity, sediment, or excess nutrients may contribute to low DO.

Critical Thinking Questions

1. Answers will vary.
2. Answers will vary. DO concentrations will vary with changes in temperature (summer to winter), flow velocity (spring floods, late summer drought), and nutrient availability (spring and summer growth, fall senescence). Dissolved oxygen is usually higher in the winter where open water is whipped by the wind.
3. Temperature affects the amount of oxygen the water can hold. Cooler water holds more oxygen than warmer. Increased turbidity may reduce the DO level. Small sediment particles would have organic materials that could be acted upon by bacteria that, in turn, use oxygen from the water.
4. Answers will vary. Reducing sediment and nutrients generally improves the dissolved oxygen levels.
5. Cool, clear streams with a moderate to high gradient are necessary for trout because they require large amounts of oxygen.

Student Activity 5.5
Analyses and Conclusions

1. Answers will vary. High BOD suggests poor water quality.
2. Answers will vary. High levels of nutrients, decomposing material, and high turbidity all contribute to high BOD. Low BOD is found in opposite situations.
3. Answers will vary but may include reducing sources of excess nutrients and turbidity.

Critical Thinking Questions

1. Answers will vary. Microorganisms might be using a large amount of the available DO, or macroorganisms might have plenty of DO available.

2. Answers will vary. In the late summer, when there is more decomposing material and the stream flow slows, BOD increases. Usually water temperature is higher in the late summer and early fall.
3. Temperature affects the percent of saturation of DO. If temperature drops, it may have a lethal combined effect with BOD on the macroorganisms. Turbidity may prevent photosynthesis by deeper plants, causing them to stop producing oxygen and begin decomposing. Turbidity would also add nutrients.

Lesson 6: Answers to Questions

Student Information 6.1
Questions

1. Coliform bacteria can be isolated through cultural techniques and used as indicators of human impact on a waterway. They are associated with other, more dangerous pathogenic bacteria whose presence is difficult to determine and dangerous to culture.
2. Typhoid, cholera, dysentery, and hepatitis.
3. Fecal coliform are introduced to water via the waste products of vertebrates.
4. No coliform should be found in drinking water; 200 colonies per 100 mL is the normal level for swimming water.
5. It has a narrow range for growth and is, therefore, easily differentiated from other bacteria colonies. It is not regarded as specifically threatening to humans, but it may be associated with dangerous bacteria.
6. Answers will vary but should mention animal and human coliform sources rather than visual appearance of the water or site.

Student Information 6.2
Questions

1. To avoid bacterial contamination of the test, and to ensure that all the colonies cultured come only from the sample.
2. Samples must be refrigerated to slow bacteria reproduction. The sample will become more dissimilar to the source as the bacteria live and reproduce in the sample jar. For this reason, the sample should be cultured as soon as possible, usually within six hours.
3. Wash hands, do not handle materials often, sterilize containers, wash table and equipment with alcohol, and keep contaminants to a minimum.
4. The Petri dish contains live bacteria. Though fecal coliform may be innocuous, other bacteria that are present may be pathogenic.
5. When making a dilution, it is important to consider the source of the water and, therefore, the level of fecal coliform likely to be present.
6. These tests are used as a control to ensure that sterile lab procedures have been followed.

Student Activity 6.3
Analyses and Conclusions

1–2. Answer will vary according to fecal coliform concentration.

3. Answers will vary by site. Large rivers often have outdated treatment facilities that may contribute to FC contamination. Farm land and feed lots may contribute bacteria from the animal waste that runs off the land. Old septic systems, pit toilets, excess nutrients, and so forth may contribute to contamination. Flooding greatly raises coliform counts.

Critical Thinking Questions

1. Answers will vary by results but may recognize the role of animal or human activities rather than whether the site looks clean or polluted.
2. Answers will vary. (Fecal coliform bacteria are not known to be harmful to other organisms but the associated bacteria may be, particularly to humans.)
3. Answers will vary. Seasonal floods often raise the FC level of streams as treatment plants are inundated and sewer lines back up. In summer, lower stream or river volume may concentrate bacteria into a smaller amount of water, raising FC levels. In winter, sewage treatment plants may not treat end-product because people do not use the river as much in the cold weather. Fecal coliform rates, thus, may go up because of untreated effluent.
4. Because the P/A test is quick and easy, it is preferred when there is little reason to expect FC contamination. It is also appropriate for testing drinking water, in which the level should be zero.

5. Answers will vary by site but may include: Control agricultural runoff, upgrade septic systems, and better evaluate and treat effluent.
6. a. Answers will vary, but water with zero fecal coliform count would be the only one safe to drink.
   b. Answers will vary but should indicate the introduction of fecal coliforms from some source. River levels could result from sewage effluent, animal sources, septic systems.
   c. Answers will vary but should be related to reducing contaminants.
   d. Boil the water, treat with a chemical, or put through a filter.

Lesson 7: Answers to Questions

Student Information 7.1
Questions

1. The permitting system allows the government to watch over the country's aquatic resources, making sure they are used in the best interest of the people; it also ensures that water quality regulations and wildlife laws are observed.
2. It also considers environment, water quality, culture, history, aesthetics, public interest, recreation, and public safety.
3. The purpose of FWPCA was to restore and maintain the chemical, physical, and biological integrity of this nation's waters.
4. The steps in an environmental assessment are: Submit a permit application, agency reviews the application and studies the site, public comment; draft assessment; and final permit decision.

Student Activity 7.2
Analyses and Conclusions

1. Answers will vary.

Critical Thinking Questions

1. Answers will vary. Determining the environmental impact is usually most time-consuming, because it involves performing and analyzing many tests, then determining probable effects of the project and comparing these to baseline information.
2. Answers will vary according to permit site and project type. Should include rationale for making decision based on guidelines.
3. Answers will vary.

Student Activity 7.3
Analyses and Conclusions

1. Answers will vary. Sometimes this assessment can be made only as a judgment, using as much objective material as is available.
2. Answers will vary.

Critical Thinking Questions

1. Answers will vary. Suggestions may include investigation of additional alternatives, mitigation activity, site changes, or proposal modification.
2. Answers will vary. Compliance does not mean a proposal is harmless. The granting agency tries to meet the needs of human development and environmental health. It is a delicate balancing act and is often a matter of pursuing or encouraging the least-damaging alternative. A dam, for instance, is always environmentally detrimental, but humans may decide they need the electricity, flood control, recreational opportunities, or water supply more.
3. Answers will vary but might refer to a proposal for activities that ensure national defense or the health and well-being of humans.

4–5. Answers will vary.

Student Activity 7.4
Analysis and Conclusions

1. Answers will vary but should include key arguments.
2. Answers will vary but may indicate that a persuasive presenter used clear, concise arguments or covered one or two important elements strongly rather than every possible argument.
3. Answers will vary but may indicate that the presenter was ill-prepared, spoke too quietly, did not face audience, or made up or omitted details.

4–5. Answers will vary.

Critical Thinking Questions

1–2. Answers will vary.

3–4. Answers will vary but may include socioeconomic studies or soil surveys.

Appendix C

Resources

Suppliers of Chemical Kits

CHEMetrics, Inc.
Route 28
Calverton, VA 22106
800-356-3072 • Fax: 703-788-4856
in VA: 703-788-9026
http://www.chemetrics.com

HACH Company
P.O. Box 389
Loveland, CO 80539-0389
800-227-4224 • Fax: 303-669-2932
303-669-3050
http://www.Hach.com

LaMotte Chemical Products Company
P.O. Box 329
Chestertown, MD 21620
800-344-3100 • Fax: 301-778-6394
in MD: 301-778-3100
http://www.Lamotte.com

Millipore Corporation

East
Bedford, MA 01730

800-645-5476
Fax: 617-275-8200
in MA: 617-275-9200
http://www.millipore.com

West
448 Grandview Drive
South San Francisco, CA 94080
800-632-2708
Fax: 415-952-1740

General Scientific Suppliers

BioQuip
17803 LaSalle Avenue
Gardena, CA 90248-3602
310-324-0620 Fax: 310-324-0620

Carolina Biological Supply Company
2700 York Road
Burlington, NC 27215
800-334-5551
Fax: 800-222-7112
http://www.carosci.com

Connecticut Valley Biological
82 Valley Road
P.O. Box 326
Southampton, MA 01073
800-628-7748 • Fax: 413-527-8286

Cynmar Corporation
131 North Broad Street
Box 530
Carlinville, IL 62626
800-223-3517 Fax: 800-754-5154
http://www.cynmar.com

Fisher Scientific
4901 West LeMoyne Street
Chicago, IL 60651
800-955-1177 • Fax: 312-378-7174
312-378-7770
http://www.fisheredu.com

Flinn Scientific, Inc.
P.O. Box 219
131 Flinn Street
Batavia, IL 60510-0219
800-452-1261 • Fax: 708-879-6962
http://www.flinnsci.com

Forestry Suppliers, Inc.
P.O. Box 8397
Jackson, MS 39284-8397
800-647-5368 • Fax: 800-543-4203

Nasco
901 Janesville Avenue
Box 901
Fort Atkinson, WI 53538-0901
800-558-9595 Fax: 414-563-8296
http://www.nascofa.com

Sargent-Welch Scientific Company
911 Commerce Court
Buffalo Grove, IL 60089-2362
800-727-4368 • Fax: 708-677-0624

Ward's
East
5100 West Henrietta Road
P.O. Box 92912
Rochester, NY 14692-9012
800-962-2660 • Fax: 800-635-8439

West
815 Fiero Lane
P.O. Box 5010
San Luis Obispo, CA 93403
800-872-7289 • Fax: 805-781-2704

Organizations and Web Sites

In addition to your state, local, and regional governmental offices, the following organizations and federal government agencies can provide useful information.

Adopt A Stream Foundation, 600 128th Street, SE, Everett, WA 98208. Phone: 206-316-8592. Water-quality monitoring, clean-up programs.

American Rivers, 801 Pennsylvania Avenue SE, Suite 400G, Washington, DC 20003-2167. Preservation and restoration of rivers. http://www.amrivers.org

American Water Resources Association, 950 Herndon Parkway, Suite 300, Herndon, VA 20170-5531. Phone: 703-904-1225. Fax: 703-904-1228. Posters and booklets. http://www.awra.org

American Water Works Association, 6666 West Quincy Avenue, Denver, CO 80235-3098. Phone: 303-794-7711. Campaign to preserve water resources. http:/www.awwa.org/

America's Clean Water Foundation, 750 First Street NE, Suite 911, Washington, DC 20002-4241. Educational materials.

Big Rivers. Web site and magazine about the Mississippi. http://www.luminet.net/~bigriver

Busy Teacher Web Site. Environmental resources on the Internet. Crosslinks to many organizations in every state. http://www.ceismc.gatech.eduBusyT/TOC.html

Center for Environmental Information, 46 Prince Street, Rochester, NY 14607. Phone: 716-271-3550. Great Lakes information. http://www.awa.com/nature/cei

Center for Marine Conservation, 1725 DeSales Street NW, Suite 500, Washington, DC 20036. Phone: 202-429-5609.

Center for the Great Lakes, 35 East Wacker Drive, Suite 1870, Chicago, IL 60601.

Classroom of the Future: Exploring the Environment. NASA program for environmental education and telecommunications. http://www.cotf.edu

EcoLinks. Comprehensive list of links to environmental information. http://www.ecomall.com

EcoNet Home Page. Environmental news, resources, and directory of organizations. http://www.econet.apc.org/econet

Eisenhower National Clearinghouse. Teacher-made materials and curriculum guides. http://www.enc.org

EnviroLink Library. Perhaps the largest online library. Many good links. http://www.envirolink.org/

Environmental Education on the Internet. Environmental resources on the Internet. Links to organizations in every state. http://eelink.umich.edu/

Environmental Organization Web Directory. Links to many groups on river and water resources. http://www.webdirectory.com/Water_Resources/Rivers

Globe Project. International monitoring by students. http://www.globe.gov/

GREEN (Global Rivers Environmental Education Network). Long-time site for student monitoring. http://www.igc.apc.org/green/

Greenpeace USA, 1436 U Street NW, Washington, DC 20009. Phone: 202-462-1177. http://www.greenpeace.org

HLM Mathematics, Science, and Technology Center in Michigan. Fantastic home page with water resources and many other connections. http:imc.lisd.k12.mi.us

International River Network. Dedicated to helping people organize to protect and restore rivers of the world. http://www.irn.org/

Izaak Walton League of America, 1401 Wilson Boulevard, Level B, Arlington, VA 22209-2318. "Save Our Streams" program, publications. http://www.iwla.org

Kentucky Water Watch and Student Testing Programs. Student sites for river monitoring. http://www.state.ky.us/agencies/nrepc/water/

Macroinvertebrate Key. Online identification key for benthic macroinvertebrates. http://monticello.avenue.gen.va.us/Community/Environ/EnvironEdCenter/Habitat/StreamStudy

Mississippi River Basin Alliance. Umbrella for grassroots organizations committed to nurturing the Mississippi River basin. http://www.mrba.org/mrba/

National Audubon Society, National Education Office, Route 4, Box 171, Sharon, CT 06069. http://www.audubon.org

National Climatic Data Center. http://www.ncdc.noaa.gov

National Geographic Society, 1145 17th Street NW, Washington, DC 20036. Publishes *National Geographic*, *National Geographic World,* and *National Geographic Traveler*. http://www.nationalgeo.com

National Wildlife Federation, 1400 16th Street, NW, Washington, DC 20036-2266. Phone: 202-797-6800. http://www.nwf.org

Natural Resources Defense Council, Public Information, 40 West 20th Street, New York, NY 10011. Phone: 212-727-2700. http://nrdc.org

Nature Conservancy, The, 1815 North Lynn Street, Arlington, VA 22209. Phone: 800-628-6860. http://www.tnc.org

Plan-It Earth, Illinois Department of Natural Resources. Environmental resources on the Internet. Links to organizations in every state. http://www dnr.state.il.us/nredu/classrm/classrm.html

River Network. Dedicated to helping people organize to protect and restore rivers and watersheds. http://www.rivernetwork.org/~rivernet

Rivers Project, Box 2222. Southern Illinois University, Edwardsville, IL 62026-2222. Phone: 618-692-3788. For student data exchange and teacher materials. http:www/siue.edu/OSME/river

Sierra Club, Information Center, 730 Polk Street, San Francisco, CA 94109. Phone: 415-776-2211. http://www.sierraclub.org

Soil and Water Conservation Society. Soil, water, and related natural resources. Many links. http://www.swcs.org

U.S. Army Corps of Engineers. http://www.usace.mil/env.html

U.S. Environmental Protection Agency, 401 M Street SW, Washington, DC 20460. EPA Wetlands Information Hotline: 800-832-7828. Mon.–Fri. 9:00 A.M.–5:30 P.M. EST.

—Software for Environmental Awareness. http://www.epa.gov/grtlakes/seahome

—Surf Your Watershed. http://www.epa.gov/surf

—Volunteer Monitoring Program. http://www.epa.gov/OWOW/monitoring/vol.html

U.S. Fish and Wildlife Service, Department of the Interior, Washington, DC 20240. Information on wetlands. http:www.fws.gov

U.S. Geological Survey (USGS), National Water Information Clearinghouse, 423 National Center, Reston, VA 22092-0002. Phone: 800-426-9000. http://www.usgs.gov

—Biological Resources Division. National Biological Information Infrastructure (NBII) gives biological information and networks in all the states. http://www.nbs.gov

—Map Sales, Federal Center, Box 25286, Denver, CO 80225. Phone: 303-236-7477. Map-ordering information.

—National Water Quality Assessment Program: Deputy Assistant Chief Hydrologist, NAWQA Program, U.S. Geological Survey, National Center, 12201 Sunrise Valley Drive, MS 413, Reston, VA 22092. http://www.rvares.er.usgs/gov/nawqa

—Public Inquiries Office, 503 National Center, Room 1-C-402, 12201 Sunrise Valley Drive, Reston, VA 22092. Phone: 703-648-6892; 800-USA-MAPS.

U.S. Public Interest Research Group (PIRG). Provides research and information on environmental issues. http://www.pirg.org

Water Environment Federation, 601 Wythe Street, Alexandria, VA 22314-1994. http://www.weg.org

West Virginia K–12 RuralNet Project. Student sites for river monitoring. http://www.wvu.edu/~ruralnet

Wilderness Society, The, 1900 17th Street, Washington, DC 20006. Phone: 202-833-2300.

Zebra Mussel Information Resources. Lots of information on the zebra mussel in American waters. http://www.nfrcg.gov:80/zebra.mussel

Books and Classroom Kits

Andrews, William A. *Investigating Aquatic Ecosystems.* Scarborough, Ontario: Prentice-Hall, 1987.

American Public Health Association (APHA). *Standard Methods for the Examination of Water and Wastewater*, 17th Ed. Washington, DC: APHA, American Water Works Association, and Water Pollution Control Federation, 1989.

Campbell, Gayla and S. Wildberger. *The Monitor's Handbook.* Chestertown, MD. The LaMotte Company, 1992.

Cauduto, Michael. *Pond and Brook—A Guide to Nature Study in Freshwater Environments*, 2d Ed. New Jersey: Prentice-Hall, 1985.

Cole, Frank R. *The Flies of Western North America*. Berkeley, CA.: University of California Press, 1969.

Herbert, Paul A. *Great Lakes Nature Guide*. Lansing, MI: Michigan United Conservation Clubs, 1972.

Izaak Walton League. *Save Our Streams*. Arlington, VA: Izaak Walton League of America.

Kopec, J. and S. Lewis. *Stream Quality Monitoring: A Citizen Action Program*. Columbus, OH: Ohio Department of Natural Resources, 1990.

Lehmkuhl, Dennis. *How to Know the Aquatic Insects*. Dubuque, IA: W. C. Brown Co., 1979.

Lopinot, A. C. *Aquatic Weeds: Their Identification and Methods of Control*. Fisheries Bulletin #4. Springfield, IL: Department of Conservation, Division of Fisheries, 1971.

McCafferty, P. W. *Aquatic Entomology: The Fishermen's and Ecologist's Guide to Insects and Their Relatives*. Boston: Jones and Bartlett, 1981.

McDonald, B., W. Borden, and C. Lathrop. *Citizen Stream Monitoring: A Manual for Illinois*. Springfield, IL: Illinois Department of Energy and Natural Resources, 1990.

Merritt, T. W. and K. W. Cummins. *An Introduction to the Aquatic Insects of North America*. Dubuque, IA: Kendell/Hunt, 1996.

Mitchell, Mark K. and William B. Stapp. *Field Manual for Water Quality Monitoring*, 4th Ed. Dexter, MI: Thomson-Shore, Inc., 1990.

Murdoch, Tom, K. O'Laughlin, and M. Cheo. *The Streamkeeper's Field Guide*. Everett, WA: The Adopt-A-Stream Foundation, 1994.

Needham, James G., and Paul R. Needham. *A Guide to the Study of Freshwater Biology*. San Francisco: Holden-Day, 1962.

Palmer, C. M. *Algae and Water Pollution*. Cincinnati: USEPA Office of Research and Development. EPA-600/9-77-036, 1977.

Pennak, Robert. *Freshwater Invertebrates of the United States*. New York: John Wiley & Sons, 1978.

Reid, George K. *Pond Live—A Golden Guide*. New York: Golden Press, 1967.

Richards, Carl, D. Swisher, & F. Arbona, Jr. *Stoneflies*. New York: Benn Brothers, 1980.

RiverWatch Network. *Guide to Macroinvertebrate Sampling*. Montpelier, VT: RiverWatch Network, 1993.

Sloat, Sharon and Carol Ziel. *The Use of Indicator Organisms to Assess Public Water Safety*. Loveland, CO: Hach Company, 1992.

Tennessee Valley Authority. *Common Aquatic Flora and Fauna of the Tennessee Valley,* Water Quality Series Booklet #4. Chattanooga, TN: Tennessee Valley Authority, 1994.

Terrell, Charles R. and Patricia Bytnar Perfetti. *Water Quality Indicators Guide: Surface Waters*. U.S. Department of Agriculture Soil Conservation Service, September 1989.

Yates, Steve. *Adopting a Stream: A Northwest Handbook.* Everett, WA: The Adopt-A-Stream Foundation, 1988.

Glossary

abiotic (ay BY ah tik) not biological.

adaptation the process of change in the physical or behavioral traits of a species due to some environmental pressure; a chance event that occurs, thus allowing certain organisms to survive at a better rate due to some current advantage they have in a particular physical or behavioral trait.

aerobic organism (er OH bik OR guh niz uhm) an organism that requires oxygen.

aerobic oxidation (er OH bik AHX ih day shun) the chemical process that, when microorganisms feed on organic matter, breaks down that organic matter, reducing the supply of dissolved oxygen gas in the water.

aesthetic (es THE tik) **value** in terms of habitat assessment, perception of what constitutes a desirable body of water. Based on individual interpretations, which may take into account visual appearance, smell, feel, and sounds, even taste if the water is used for drinking.

algae (AL jee) one-celled plants.

algal (AL juhl) **bloom** abnormally rapid algae growth, often caused by high temperature or high levels of organic materials.

anaerobic bacteria (an er OH bik bak TEER ee ah) bacteria that do not require oxygen; many survive only in low-oxygen environments.

aquatic (ah KWAH tik) **biologist** a scientist who studies freshwater organisms and their environments.

aquatic (ah KWAH tik) **habitat** water-oriented setting that contains a variety of flora (plants) and fauna (animals), some seen and some unseen.

bacteria (bak TEER ee ah) the plural of *bacterium*.

bacterium (bak TEER ee uhm) one-celled microorganism that has no chlorophyll.

balanced ecosystem a closed ecosystem that is functioning properly.

bank erosion the condition in which soil detaches from the bank and moves into the stream.

bedrock the solid expanse of material that forms a layer of rocks below unconsolidated surface materials such as soil; sometimes it is exposed to the surface.

benthic macroinvertebrate (BEN thik MAK roh ihn VERR te brayt) bottom-dwelling invertebrate (without backbone) organism visible without the aid of a microscope.

benthic macroinvertebrate (BEN thik MAK roh ihn VERR teh brayt) **pollution tolerance index** an index that places benthic macroinvertebrates in groups according to their level of pollution tolerance, with the most pollution-tolerant weighted most heavily, and whose results provide a rapid means of assessing stream quality.

bioassay (BY oh ASS ay) a test that provides a measure of some biological matter, not by measuring the amount of matter directly but by measuring the effect of that matter on biological organisms.

biochemical oxygen demand (BOD) the requirement, or demand, for oxygen that organic and chemical matter places on water.

biotic (BY ah tik) having to do with biology.

BOD see *biochemical oxygen demand.*

boulder sediment having greater size than 25 cm (10 inches) in diameter.

carnivore an animal that eats other animals.

channel the portion of the stream bed in which the deepest, fastest portion of the river or stream flows.

channel capacity the maximum volume of water that can pass down the river or stream at a given time.

Clean Water Act the revised Federal Water Pollution Control Act of 1948, as amended and renamed in 1977, that regulates permits for activities potential harmful to bodies of water, with the goal of restoring and maintaining the chemical, physical, and biological integrity of U.S. waters.

closed ecosystem an ecosystem sealed off from the outside world, so no materials can enter or leave; contains just enough plants to provide food and oxygen for the animals.

cobble rock from 5 to 25 centimeters (2 to 10 inches) in diameter.

coliform (KOH luh form) bacteria that propagate in the digestive systems of humans and other warm-blooded animals. See also *fecal coliform.*

colony a group of organisms, such as a group of bacteria that result from the division of a single bacterium.

condensation (kahn dehn SAY shuhn) the process in which, as water vapor cools, it releases energy and changes into a liquid state.

consumer an animal that feeds directly on plants.

control a substance whose characteristics are already known.

control plate a Petri dish that contains a sterile sample.

cost-benefit analysis a study that compares the economic gains and losses of a project or activity, sometimes with those of other alternatives.

cultural factor any human action that impacts a river or stream.

culture to propagate organisms, such as bacteria, in a specially prepared environment that enhances growth.

decomposer an organism that eats and digests dead plant or animal matter, converting it into a nutrient form available to other organisms.

density the number of organisms of one taxon in a sample or area.

detritus (duh TRY tuhs) decomposing organic matter.

dilution (di LOO shuhn) the process of reducing the concentration of matter per unit volume.

dip net any of several types of long-handled net for collecting from stream bottoms. See also *D-net* and *triangular net*.

disinfectant substance that inhibits or destroys microorganisms.

dissolved oxygen (DO) elemental oxygen (O_2) gas in a solution of water.

diversity index an index that represents the number of species or types of organisms present in a biological community, in a system in which a high index value is an indicator of higher water quality.

D-net a long-handled dip net, with netting in D shape, for collecting from stream bottoms.

dredged material material removed from an underwater site (such as a river or bay bottom).

ecology the study of the interactions among plants, animals, and their nonliving environment within an ecosystem.

ecosystem a system of interactions between living organisms and their environment.

effluent the watery materials resulting from sewage treatment.

Elodea (ehl OH dee ah) a leafy, submerged aquatic plant.

embeddedness description of how much of the surface area of the larger materials in a channel is covered by sediment.

endangered in immediate risk of extinction.

energy fuel or capacity for doing activity.

enterobacteria (enh TEHR oh bak TEER ey ah) rod-shaped, gram-negative (meaning it does not respond to a test known as Gram's bacterial stain), and non-spore-forming bacteria; the family of bacteria to which coliform bacteria belong.

environment all the conditions and circumstances surrounding and affecting an organism or group of organisms.

environmental assessment a procedure for determining the effects of a proposed activity on the environment (including the human environment).

erosion (ih ROH zhuhn) in terms of stream banks, the separation of soil and rock from the bank, moving into the stream; more generally, the separation of soil and rocks from another surface by water, ice, wind, or gravity.

erosion potential the ability of soil to detach or move into the stream, as influenced by steepness, composition, and vegetation of the land.

estuarine (EHS chew reen) **wetlands** an area of land, where fresh water and salt water mix, which is covered by a shallow layer of water for some time during the year, such as at the mouths of rivers that flow into the ocean.

eutrophication (yoo troh fuh KAY shuhn) the overgrowth of one taxon in a sample or area of aquatic vegetation followed by death, decay, oxygen depletion, and an imbalance of plants and animals in the water.

evaporation (ih vap up RAY shuhn) the process in which, as water absorbs the sun's energy, it vaporizes.

evenness the number of individuals in each taxon relative to the total number of individuals in a sample of organisms; shows how uniformly individuals are distributed among taxa; in the diversity equation, expressed as P.

fauna (FAH nah) animals.

fecal coliform (FEE kuhl KOH luh form) any of a group of rod-shaped, waterborne bacteria that propagate in the digestive systems of humans and other warm-blooded animals.

field biology a study of living things outside, or in the field.

fill material material added to or removed from a land site, typically a construction site.

fish kill the rapid death of fish in an area when algal bloom or other factors cause a significant, rapid depletion of oxygen from the water.

flora plants.

flow the volume of water passing a particular site in a particular amount of time.

flow rate the speed with which water is flowing; velocity, expressed in meters per second (m/s).

food chain the sequence of events in which energy is passed from organism to organism in the form of nutrients.

food web overlapping and intersecting of food chains.

fresh water water with a low salt content.

fungi (FUHN guy) the plural of *fungus*.

fungus (FUHN guhs) mushroom-like organism; a classification of plantlike organisms that have no roots, flowers, leaves, or chlorophyll. Fungi obtain nutrients from living or dead organisms and reproduce by spores.

gradient (GRAYD ee uhnt) the rate of descent of a stream; steepness.

gravel rock from 3 millimeters to 5 centimeters (0.1 to 2 inches) in diameter.

groundwater water found below the surface of the earth; subsurface water.

habitat the area or type of environment in which an organism or ecological community lives.

habitat assessment a survey of a particular site, ranking its suitability as an environment for plants and animals.

herbivore (EHR bih vohr) see *primary consumer.*

hydrologic (hy druh LAHJ ihk) **cycle** the endless cycling or interchange of water among the oceans, the lower atmosphere, the land surface, and underwater reserves.

hydrologic (hi druh LAHJ ihk) **modification** a change in the course of water at any point in the hydrologic cycle.

incubation keeping testing material in an environment of controlled temperature in order to provide optimal conditions for growth and development.

index a rating system that assigns a value to an object or process, or to specific qualities it may possess.

indicator organism an organism that indicates other events or the activity of other organisms.

indices (IHN duh seez) the plural of *index.*

infiltration the process in which surface water seeps into the earth.

invertebrate (ihn VERR te brayt) an organism without a backbone.

kick net a seinelike net used to capture benthic macroinvertebrates and other small animals when the substrate upstream is disturbed by foot action.

lacustrine (LA kuh strinn) **wetlands** an area of land associated with a pond, lake, or reservoir that has slow or low movement of water, and which is covered by such water for some time during the year and has saturated soil from that contact.

levee (LEH vee) a low ridge that keeps water within a river channel.

limiting factor a variable, either natural or human-made, that limits conditions or actions.

lower bank the portion of the stream bank that comprises the area from the normal high-water line to the low-water line.

macroinvertebrate (MAK roh ihn VERR teh brayt) an invertebrate (without backbone) organism visible without the aid of a microscope.

macrophyte (MAK roh fyt) a larger plant organism.

marine biologist a scientist who studies saltwater organisms and their environments.

marine wetlands an area of land near an ocean, covered by a shallow layer of salt water for some time during the year, such as tidal pools, mud flats, and beach areas.

material safety data sheet (MSDS) printed material that describes the proper safety precautions to observe when using specific chemicals.

medium a nutrient-balanced substance on which microorganisms are grown.

membrane filtration (MF) a test to determine the abundance of fecal coliform in water, which involves vacuuming a sample through a filter, incubating the filter at a temperature at which nonfecal coliform cannot survive, then counting the number of resulting colonies.

metabolism all the chemical activities that take place in an organism.

microinvertebrate (MIH kroh ihn VERR teh brayt) an invertebrate (without backbone) organism so small that seeing it requires a microscope.

migration seasonal movement of animals.

mitigation in environmental management, some compensatory action, often in another location, to offset the effects of an intended action.

National Pollution Discharge Elimination System (NPDES) a set of regulations covering industrial and municipal source discharges of chemical, biological, and thermal pollutants into U.S. waters.

naturalness the apparent status of an environment as untouched by human influence.

navigable (NAV ih guh bull) **capacity** the accessibility of a body of water for travel by ship or boat.

navigation vessel traffic.

nonpoint source pollution human-made factors in a watershed other than siltation that add materials to the water that affect its suitability as habitat, specific urban, industrial, and agricultural runoff

nutrient food for an organism to grow.

omnivore an animal that eats both plants and animals. See also *scavenger.*

open ecosystem an ecosystem that is open to the outside world, allowing materials to enter or leave.

organic matter anything that is or was alive.

organism any living entity.

overall Water Quality Index (WQI) total value of water quality, obtained by adding the weighted percentage scores of nine water-quality tests.

palustrine (PAL uh strinn) **wetlands** an area of land made up of the bogs, marshes, and wet meadows found in forests and prairies, covered by a shallow layer of water for some time during the year and with saturated soils from that contact.

parameter (puh RA muh ter) specific measurable aspect of some situation or other phenomenon; in terms of habitat assessment, some aspect of the stream environment.

pathogenic (path uh JEHN ihk) disease-causing.

percent saturation the percent of milligrams of oxygen gas dissolved in one liter of water at a given temperature compared with the maximum milligrams of oxygen gas that can dissolve in one liter of water at the same temperature.

periphyton (PEHR ih fy tuhn) an organism attached to underwater surfaces, such as a sponge or a rotifer.

permit a document from a governing agency that grants permission to carry out some activity, in particular activities that may have ecological, social, or economic impacts.

phosphate buffer a phosphate compound dissolved in water, similar to bacteria's natural environment.

photosynthesis the process of using sunlight to convert carbon dioxide and water into food energy, characteristic of all chlorophyll-possessing organisms.

phytoplankton (FY toh plank tuhn) a type of plankton that is a photosynthetic plant.

plankton a microscopic floating or weakly swimming organism. See also *phytoplankton* and *zooplankton*.

point-source discharge discharge whose origin can be traced to a particular source.

pool in a river or stream, an area of slow, quiet water that spreads out.

potable (POH tah buhl) **water** water that is safe to drink.

precipitation (prih sihp uh TAY shuhn) the deposits of water that reach Earth's surface from the atmosphere.

Presence/Absence (P/A) test a simple screening test for coliform bacteria in a water sample, in which a water sample is mixed with testing material, then incubated; if the sample changes color, coliform bacteria are present.

primary consumer an animal that does not consume other animals, just plants, and is, in turn, consumed by other animals; also called *herbivore* or *producer*.

producer an organism that produces its own food, some of which will be passed along the chain when the organism is eaten or released when decomposed.

protist plant or plantlike one-celled organism.

protocol (PRO tuh kohl) a uniform procedure for the application and significance of a test or survey, such as for an index.

public interest review a review that weighs the comments and concerns of the public along with scientific studies and analyses relating to some proposal that requires a government permit.

Q-value percent common unit used for comparing the results of different kinds of water-quality tests.

respiration the exchange of gases in order to obtain the compounds required for energy, specifically, using oxygen, which organisms use to break down food materials into usable materials plus carbon dioxide and water.

richness the number of taxa in a sample of surveyed organisms.

riffle a relatively shallow area of a river or stream over which water flows unevenly.

river biologist a scientist who studies the organisms of rivers, ponds, lakes, and the surrounding wetlands.

riverine wetlands an area of land associated with the moving water of a river or stream, which is covered by a shallow layer of water for some time during the year and has saturated soil from that contact.

run an area of moving, relatively deeper water, whose surface may be choppy.

runoff water on the Earth's surface that flows across the land surface to streams.

saline salt-containing.

scavenger an animal that will eat an animal that is already dead; sometimes used to refer to an omnivore; also a garbage eater.

secondary consumer an animal that eats primary consumers.

Section 404 the section of the Federal Water Pollution Control Act that established the U.S. Army Corps of Engineers permit program that regulates physical changes to wetlands.

sensory perception the increase in awareness via sight, hearing, smell, touch, or taste.

sewage human waste material carried in water through sewer pipe systems.

siltation soil being carried into a stream or river and deposited there.

sterile lacking any form of life, even microscopic.

sterilization (sterhr uh luh ZAY shuhn) a process of creating an environment free from unwanted microorganisms.

stream bank the land adjoining a stream, from break in the general slope of the surrounding land to low water line.

substrate (SUB strayt) materials found at the bottom of the stream bed, on which many aquatic organisms live or reproduce.

surface water fresh water on the surface of the Earth, such as lakes, rivers, and streams.

taxa the plural of taxon.

taxon a category into which related organisms are placed, such as species or apparent species.

tertiary (TUHR shee ayhr ee) **consumer** an animal that eats carnivorous animals.

threatened declining in number throughout a species range.

titration (ty TRAY shun) the addition of a liquid to another liquid drop by drop.

tolerance in field biology, an organism's ability to exist in the presence of some specific relevant factor, such as pollution.

too numerous to count (TNC) a condition in which a sample contains so many colonies that they are difficult to count.

topographic (tahp uh GRAF ihk) **map** a map with a sufficiently large scale to show detailed natural and cultural surface features.

trait any characteristic that can be passed from parent to offspring.

transpiration (trans puh RAY shuhn) the evaporation of water from plants.

triangular net a long-handled dip net, with netting in triangular shape, for collecting from stream bottoms.

turbidity (ter BIHD ih tee) the degree of lack of clarity in a liquid or gas.

upper bank the portion of the stream bank that comprises the land area from the break in the general slope of the surrounding land to the normal high-water line.

vegetated covered with plants.

water cycle see *hydrologic cycle.*

watershed the area drained by a river system; the land area from which water flows toward a common stream in a natural basin.

weighting factor the value or points given to a factor in an index to make it count for more in the quantitative analysis.

wetlands an area where the water table is at, near, or above the land surface long enough during the year to support the growth of water-dependent vegetation.

wetlands map an overlay for a topographic map that delineates and identify types of wetland areas.

Winkler method a process for determining the amount of dissolved oxygen in a water sample by adding one chemical to that sample to form an insoluble colored precipitate, then counting the number of drops of a second chemical added until the solution becomes colorless.

zooplankton (ZOH uh plank tuhn) a type of plankton that is an animal.